RIDING ON RUBBER

Cover: The Kingston Mill with The Hall behind. The River Avon is in the foreground.
The unusual octagonal building shown is known today as 'The Dovecote' but
was built as a wool drying stove. Painted by Elizabeth Tackle around 1850.

At the Beginning

Beneath the shadow of a stately Elizabethan mansion at Bradford-on-Avon in the heart of the West country stands a small but compact factory on the banks of a quietly flowing river.

Since the days of Elizabeth with her merchant adventurers and their commercial expansion, Britain has made a glorious journey of pioneering enterprise along the never ending path of progress.

It has been a long and triumphal journey from the days of Elizabeth to the new age of rockets and radar, jet propulsion and tanks, flame-throwers and mines.

Reproduced from *Six Vital Years* published in 1946 by George Spencer, Moulton & Co. (GSM)

Riding on Rubber

The Story of Bradford on Avon's world-renowned Rubber Industry

Dan Farrell

Ex Libris Press in association with Bradford-on-Avon Museum

Published in 2017 by
Ex Libris Press
11 Regents Place
Bradford-on-Avon
Wiltshire   BA15 1ED

in assocation with
Bradford-on-Avon Museum

ISBN 978-1-912020-66-9

Origination by
Ex Libris Press
Bradford-on-Avon

www.ex-librisbooks.co.uk

Typeset in 10.5/14 point Souvenir

Printed by TPM Ltd.
Farrington Gurney, Somerset

Contents

	Introduction	7
1	Origin	9
2	The English Project	13
3	Establishment and Expansion	18
4	Setback and Resurgence	27
5	Mixing and Rolling	38
6	Honour Vindicated	48
7	Post-War Ambition	61
8	Hitting the Buffers	72
9	Avon in Bradford	75
10	Moulding and Modernization	84
	Epilogue	95
	Epitaph	98
	The Rubber Industry's buildings: Kingston House, Greenland Mills, Kingston Mill, Manvers House, New Mills and the Vaults, Lamb Building, Centenary Building, Abbey Mill, Grist Mill, The Hall	99
	Appendix: Charles Goodyear	109
	Bibliography	111
	Acknowledgements; About the author	112

Introduction

It is difficult to imagine the modern world without rubber. It is used for a myriad of different products from the soles of our shoes to the swimming caps on our heads. Cars, trains, aeroplanes and ships are stuffed with rubber hoses, seals, bushes and springs. Electrical devices rely on rubber for insulation; consumer products use it for sealing, texture and tactility. Clothing is improved by elastic waistbands and cuffs, by stretch fabrics and waterproofing. The list is endless – this seemingly simple material forms a fundamental part of our everyday lives. Yet before the mid-19th century, rubber in any practical long–lasting form simply did not exist.

This book, aside from giving an insight into the origins of rubber and the manifold difficulties in transforming the sticky latex tapped from rubber trees into a durable 'synthetic leather', seeks to detail how the small west–country weaving town of Bradford on Avon came to be so central to the British rubber industry and became the pre–eminent source of rubber railway mechanicals to the whole world. It does not attempt to be an academic text or a technical handbook, but rather an informative – and hopefully lively – account of this remarkable tale of courage, determination and innovation.

The rubber industry in Bradford on Avon is indelibly linked to two companies – the Moulton Rubber Company and their successors, George Spencer, Moulton and Co. It has been said that Stephen Moulton is the 'forgotten man' in the British rubber story; one hopes that those who read this book will gain an appreciation of how this man's belief in the potential of vulcanized rubber led him to take great risks in the name of progress. Certainly the rubber and railway industries benefitted greatly from his conviction and Bradford on Avon also did, as his company provided living wages and livelihoods – directly and indirectly – to the vast majority of the working population of the town for nearly 150 years. It is to them that this book is dedicated.

Rubber tapper moulded in stone on the Spencer Moulton centenary
building in Kingston Road (DF)

1　Origin

"There is probably no other inert substance which so excites the mind"

Charles Goodyear

Whilst the Mayan and Aztec civilizations of South America had some knowledge of what we know today as 'rubber', the Olmecs (whose name means 'rubber people') used 'rubber' balls for the Mesoamerican ball game as far back as 1500 BC. It is perhaps expedient for us to move closer to modern times. By the sixteenth century, across large areas of equatorial America the native population were familiar with the process of obtaining a white juice (known today as latex) from certain trees. An incision from an axe or a knife and the liquid would flow, with this liquid used to coat − waterproof − fabrics and leathers; or to be pasted over clay formers to make rubber containers.

The Indians in Quito called the material '*caoutchouc*' from the native '*caa ochu*' − ''the tree that weeps'. In the language of the Incas, '*cauchu*' translates as 'he who casts the evil eye' − perhaps symbolic of the seeming magical powers of the tears of the tree. The trees themselves − there are several which can be tapped for latex − include *Hevea brasiliensis*, *Castilla elastica* and *Ficus elastica* (the latter being found predominantly in Malaysia). The term − and henceforth popular name for this new material − 'India rubber' was coined in the 1770s by English philosopher and chemist Joseph Priestley. "I have seen a substance excellently adapted to the purpose of wiping from paper the mark of black lead pencil" he wrote, having obtained a small cube of 'vegetable gum' from stationers Nairne's of London. On learning of its origin, he combined this with the 'rubbing' action needed to erase pencil marks to give it the label of 'India rubber'. The learned men of the scientific world preferred the more classical 'caoutchouc' for another century before finally adopting the layman's expression.

In the early nineteenth century, rubber was being used for the manufacture of clothes, shoes and many other items − in a similar, though rather more sophisticated, way to the sixteenth century Americans. It was particularly successfully used in England, where the somewhat temperate climate did not expose rubber goods to extremes of heat that made it sticky and shapeless, or to bitter cold that made it brittle and prone to cracking.

A number of notable rubber pioneers emerged in England − chief amongst these was Thomas Hancock. Born in 1786 in Marlborough, Wiltshire, Hancock set up a coachbuilding business in London with his brother John sometime around 1815. Wishing to protect passengers in his coaches from the weather, he began experiments in waterproof fabrics using rubber dissolved in turpentine. This proved fruitless but he appreciated the elasticity of India-rubber and turned to how this could improve the fit of articles of clothing – particularly gloves but also 'any article of dress where elasticity is desirable at any particular part' – and these 'elastic strips' were patented by Hancock in 1820.

Many of Hancock's elastic strips failed whilst being sewn into clothing or even whilst being cut from raw material. With his (at that time) limited knowledge of the properties of rubber, Hancock experimented further and learned that heat could be used to re-join broken elastic strips, and rubber off-cuts could be re-cast into solid rubber blocks. He further realized that his raw material supply was far from consistent, and his new moulded blocks were rather more so. The result of this was the first mechanized aid for processing rubber – Hancock called it a 'pickling machine' in an attempt to disguise its true function. Raw rubber was cut up into small pieces and dropped into the machine, the handle was turned and after half an hour or so (depending on the quantity of rubber and the vigour of the operator) the rubber pieces were fused into one uniform lump of warm rubber. Later, this would be known as 'mastication'.

Concurrently, Charles McIntosh was also experimenting with the use of rubber to waterproof cloth. He looked to produce a 'rubber solution' by dissolving raw rubber in a suitable solvent, and using this to coat fabrics. After several false starts, great success was had by the use of coal tar oil (a by-product in the making of Coal Gas) as a solvent. McIntosh placed this solution between two sheets of fabric to avoid the sticky or brittle surface and what we know as 'oilskins' or 'Mackintoshes' were born.

By the 1830s, Hancock and McIntosh had formed an alliance. The products were of good quality, and represented the biggest and the best in the British rubber industry. But – and this was a serious drawback – they still could not handle extremes of temperature. As before, cold weather made them stiff and prone to cracking; heat made them sticky and shapeless. McIntosh, of Glasgow stock, may have had less of a problem with the latter in his early work. Hancock applied himself diligently to this issue for many years, to no avail – eventually resigning himself to reducing the effects of extremes of temperature rather than eliminating them entirely.

Meanwhile, America was in the grip of 'Rubber Fever'. Many companies sprang up to make clothing, footwear and other articles from this new wonder material from the south. Possessing perhaps more familiarity with the material but less knowledge than Hancock and McIntosh, and dealing with greater extremes of temperature, these companies had limited success. The products looked good when first made – and sold well – but soon disintegrated into a sticky, malodorous mess.

The major player in America – some say 'America's first India-rubber company' – was the Roxbury India Rubber Co. of Roxbury, Massachusetts. In 1836 Roxbury filed a patent for a 'mill and spreading machine', invented by one Edwin Chaffee. Whilst rubber had been formed into sheets using rollers for several years, the rollers on Chaffee's 'Calender' machine were steam-heated to 200 degrees F. The uniformity and strength of the rollers – and the frame that housed them – enabled the production of a very thin film of rubber and simultaneously to force this film into a cloth to make a waterproof sheet. Of great advantage was the fact that this could be done without the use of solvents and indeed the manual spreading methods used by McIntosh. The first machine had rolls of 6 ft (1.9 metres) length and weighed thirty tons.

However, despite these advances, the essential issue remained unsolved. Rubber was not durable. It melted in the heat, it cracked in the cold. It was sticky, smelly and perished easily. By the end of 1837 the American rubber boom was over. Most factories closed down and their owners and workers moved on to pastures new.

One man was not prepared to let this situation persist. He first took notice of rubber as a material in 1834, when he visited one of Roxbury's retail stores in New York. At the time, Roxbury were selling rubber lifejackets. He was unimpressed by the valve used to retain the air in these jackets and, having procured one, went away to design an improved version. Months later he returned with his new valve, only to be shown row upon row of perished rubber lifejackets. Roxbury were no longer interested in valves – they would be lucky to stay in business at all. He looked at the rubber material with a new fascination and instantly saw the possibilities, if only it could be made to be durable. He would later declare that "there is no other inert substance that so excites the mind". His name was Charles Goodyear.

Goodyear was born in 1800. His father ran a hardware business, and Charles founded his own similar company in 1826. It did not flourish, and by 1831 he was bankrupt and lost both his own business and any rights to his father's. On his release from debtor's gaol, he set himself up as an 'inventor' – hence his involvement with Roxbury and the air valves. Following that disappointment, he applied himself to the improvement of rubber like a man possessed. His first breakthrough was in 1837, with the discovery that an application of dilute nitric acid 'dried' the rubber and removed the undesirable 'tackiness'. This was better rubber than had ever been seen before, and Goodyear won a contract to supply mailbags. With the items made, he left them in storage and he took his family on a well-deserved holiday. On his return, he found them in a perished, sticky, useless state. Goodyear was distraught yet defiant. His colleagues and friends all urged him to give up his rubber project – "Rubber is over in America" was the constant refrain. He refused to accept this and insisted that "I am the man to bring it back".

Goodyear's breakthrough came in 1838. Nathaniel Hayward, a foreman at Goodyear's factory, suggested using sulphur as a compounding ingredient as a 'curing' or drying agent. This looked promising and, on Goodyear's request, Hayward patented this process and Goodyear bought the patent from him. The sulphur dried the surface of the rubber, but the fundamental problem remained. Later, by accident or design, Goodyear was to leave a sulphur-treated rubber sample on a hot stove and was astonished to find that, rather than melting, it darkened and hardened into a 'fireproof gum', 'elastic metal' or 'vegetable leather'. In his own words he stated that "the effect of this process is not simply the improvement of a substance; but it amounts, in fact, to the production of a new material. The durability imparted to gum-elastic by the vulcanising process, not only improves it for its own peculiar and legitimate uses, but also renders it a fit substitute for a variety of other substances where its use had not before been contemplated".

Whilst being fully aware that he had made the vital discovery, Goodyear could not have known what the mechanism of change in the rubber material was. 'Dr' Sam Pickles, chief Chemist at George Spencer, Moulton & Co. and doyen of the British rubber industry, would later define this change as "the result of the physical action of sulphur upon rubber"; Pickles also discovered the true structure of rubber and the role of sulphur acting as binding agent – chemically cross-linking the long polymer strands in natural rubber and giving the vulcanized rubber a permanence of structure and shape. The term 'Vulcanization' – after Vulcan, the Roman God of Fire – was coined by William Brockenden, a Director of Chas. McIntosh & Co. The importance of this discovery cannot be over-stated – it has been described as "the greatest industrial secret of the nineteenth century".

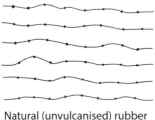
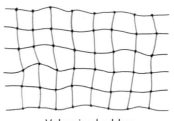

Natural (unvulcanised) rubber   Vulcanized rubber
Unstructured long chain polymer   Stabilised by sulphur cross-linking

The greatest industrial secret of the nineteenth century – how the addition of sulphur creates chemical cross-links to give natural rubber a stable, durable structure.

Goodyear's struggles were far from over. He was jailed again for unpaid debts and was unable to obtain any funding to patent his process or for further experiments – viz. to commercialize his invention. Moreover, the reputation of rubber in America was seemingly irreparably damaged – no-one of sound mind would touch it. Goodyear needed another market for his discovery, and he found a suitable ally in Stephen Moulton, a broker in New York who was a friend of William Rider, his most faithful and patient backer. And so it became that, in 1842, Moulton sailed for his home country – England – on the paddle steamer "British Queen", carrying samples of 'Goodyear's Improved Rubber' and the authority to sell the rights to the invention for the sum of £50,000 (in 2015 prices – on a relative GDP basis – around £170 million).

2 The English Project

"Your brother Richard thinks you are a fool and will bring the whole family down"

William Rider to Stephen Moulton, 1847

Stephen Moulton was born in 1794 at Whorlton, County Durham. His father, also Stephen, was a Law Stationer with offices in Chancery Lane, London. Although residing in London, the Moulton family was from the south west rather than the north east – his mother, Catherine (*née* Bellamy) was visiting her sister in Whorlton at the time of Stephen's birth.

On St Stephen's Day in 1826, Stephen Moulton married Elizabeth Hales of Williton, Somerset, at St. George's Church, Hanover Square, London. Their union was blessed with nine children – seven sons and two daughters – and blessed indeed as all survived into adulthood. By 1838, Moulton was a broker in New York and had become involved with some of the American rubber pioneers – Goodyear, Hayward and the Rider Brothers; William, John and Emory. The Riders owned a rubber factory in New York and were allies to Goodyear's cause. Moulton appeared to both Goodyear

and the Riders to be the ideal emissary to England; he knew the country, was used to brokering business deals and possessed the understanding of the importance of Goodyear's discovery but not the technical knowledge of the process.

His arrival in England in 1842 was met with coolness from the British rubber companies. They were clearly impressed by the samples of 'Improved Rubber' but suggested to Moulton that the inventor should patent the process (and hence disclose the method) so that they could form a more accurate judgment of its merits. Amongst those who examined the samples was Thomas Hancock, who – in

Stephen Moulton, 1794–1880 (AMCT/DF)

a similar manner to Goodyear's relentless pursuit – had spent the best part of twenty years attempting, without success, to improve the durability of rubber. Now, Hancock held the proof of the possibility in his hands and immediately returned to his work – "finding now that this object appeared to have been somehow or other effected, and therefore demonstrated to be practicable … I set to work in earnest, resolved, if possible, not to be outdone…". Moulton returned to New York in 1843 and urged Goodyear to progress his patent.

Hancock worked tirelessly for more than a year and, in November 1843, applied for an English patent for 'An Improvement in the Manufacture of Rubber'. Charles Goodyear filed his English patent, after much encouragement from Moulton and Rider, in January 1844. Hancock was aided by the patent laws of the time, which allowed for a preliminary filing followed by a more detailed specification within six months. The circumstances of Hancock's 'discovery' remain unclear; for his part, Hancock never denied seeing the Goodyear's samples and indeed admitted that he detected the presence of sulphur in them, but maintained that he developed his process independently. Others, including Goodyear, Moulton and Alexander Parkes (the inventor of 'cold vulcanization') took a less generous view of Hancock's behaviour and remained convinced that Hancock had 'reverse-engineered' the samples.

What cannot be argued is the fact that the first vulcanized rubber that Hancock saw came from Goodyear, and that Hancock beat Goodyear to the patent office by eight weeks. It is also worth noting that by the time he filed his complete patent in May 1844, Hancock possessed a much more detailed understanding of the vulcanization process – including how much sulphur to add and how, how to spread or mould the compound, and curing times and temperatures for different thicknesses – than Goodyear did. That said, it was to be several years before either of them could reliably produce vulcanized rubber goods.

By this time, Stephen Moulton was convinced of the commercial potential of vulcanized rubber and determined to enter the business himself. In partnership with the Rider brothers and James Thomas, an American chemist, he set out to develop an alternative vulcanization process. Moulton returned to England in 1847 (he saw this move as permanent as his family followed him – William Rider wrote to him in October – "your brother Richard thinks you are a fool and will bring the whole family down, but Mrs. M and family will leave for Liverpool in the packet ship Victoria on 25th") and in February 1847 filed a patent for rubber vulcanization using Hyposulphite of Lead in place of the free sulphur used by Goodyear and Hancock. However, the patent lacked enough detail – or those attempting to use it lacked the knowledge and skill – to be translated into production.

Relationships with Thomas soured to the point of deadlock but, undaunted, Moulton sought expert help in England. By the end of 1848 he was able to write to William Rider that "Jenny Lind (his new rubber compound, named after the Swedish opera singer) sings admirably in England…" Rider, having seen so many of Goodyear's false dawns, urged caution: "we have often been where we thought we was sure, but time proved we was wrong." This new process fell within the scope of Moulton's 1847 patent, and his next step was to prove its worth through manufacture.

Concurrent with the development of 'Jenny Lind', Moulton became aware that his American partners were perhaps less enthusiastic about the English project than he was. The Rider brothers were very much against manufacturing in England – they had enough problems of their own in New York and were well aware of the costs of setting up a rubber factory. They preferred to sell or licence the – as yet unproved – new patent, and urged Moulton to do so. Rider was not able to help Moulton with money either, citing lack of spare coin: "times arnt now what they used to was with us, and therefore we are obliged to husband every dollar we have at our command..." However, Moulton had committed himself fully to the project and there was to be no looking back.

In May 1848, following another cautionary missive from William Rider, he replied: "I must work out my own plan, at my own risk and for my own gain, exclusively." He did not delay. On the 2nd July 1848, the Chancellor of the Exchequer rose in the House of Commons to speak on the state of the nation: "I will mention here a curious circumstance, which shows the effect on the activity of trade, which is sometimes occasioned by the introduction of new descriptions of manufacture. At Bradford an abandoned cloth factory has recently been taken by a gentleman from a distance for the purpose of establishing a manufactory of articles of clothing etc. from India Rubber."

Stephen Moulton had purchased The Hall (also known as Kingston House and described as 'the best built house for the quality of a gentleman in Wiltshire'), Kingston Mill, Grist Mill, Lower Fulling Mill, a dye house, four cottages and eight acres of land for the sum total of £7,500 (in relative economic cost at 2015 prices, c. £23 million; Moulton paid £4,500 in cash and raised £3,000 by mortgage). As he had no proven vulcanization process of his own at the time, Moulton had taken a gamble that he could not afford to lose. His determination was never in doubt; indeed he had outlined his plan to Rider as below:

1 I sail from America to England
2 I take out a patent with my own cash
3 Find a capitalist
4 Investigate and get a just conclusion concerning the law of the case
5 Discover and put into practice a new and superior mode of manufacturing rubber
6 Actually make it myself in quantities and find a profitable sale for it
7 Spend a whole year of time at my own cost and risk...

Given his knowledge of the difficulties that Goodyear and Hancock had encountered in manufacturing rubber articles, item 5 appears almost breathtaking in its understatement. Moulton had set his course and would follow it resolutely; and however high the stakes were on his return to England, they were to get much higher before his success was assured.

Painting of Bradford on Avon from Westwood (AMCT/DF)

It is worth considering at this point why Moulton chose Bradford on Avon as the site for his rubber factory. It is known that Captain Palairet of Woolley Grange (on the outskirts of Bradford) purchased a quarter share of Moulton's patent for £5,000; he may well also have introduced Moulton to the district, or perhaps Moulton – with his West Country family links – had visited the town previously. Whatever the circumstance, Bradford provided what was required for a rubber manufactory: transport links (the Great West Road, the Kennet and Avon Canal and Brunel's Great Western Railway were nearby); water from the Avon for washing the raw material; sources of power – the Kingston Mill was fitted with a twenty horse-power waterwheel, and the Somerset coalfields were nearby; space to expand; and an abundant supply of labour. The railway was due to come to Bradford and to pass right by the mill property; indeed the station had already been built by 1849 but it was to be another eight years before the railway itself was in situ and operational. The view of Bradford from Westwood, painted by Elizabeth Tackle, is interesting as it is entirely contemporaneous with Moulton's arrival in the town – the railway station stands in all its glory yet there is no trace of the track.

The availability of the Kingston Mill, together with many other empty woollen mills in the district, was largely due to the disastrous failure of a local bank – Messrs.

Hobhouse, Clutterbuck, Phillot and Lowder – in 1841. The bank was heavily exposed in the town and the bankruptcies of the Staverton Mill and of Saunders and Fanner at the Church Street and (old) Abbey Mills – the largest employer in the town with over 320 hands – proved too much for them to absorb. Several of the remaining cloth manufacturers in the town closed due to loss of cash, lack of working capital or simply from poor trading conditions. Hundreds were left destitute and the workhouse was oversubscribed to the extent that three hundred of the able-bodied were put to work on 'parochial improvements' outdoors. Poor Rates rose as high as 50% – ten shillings in the pound. An emigration fund was established and many Bradfordians followed the weaving work northwards or sailed west to the New World. By 1850 the town – described as recently as 1834 as "being famous for the best manufacture of superfine woollen cloth" – had lost nearly a quarter of its residents and most of its industry. Spirits were lifted when Moulton arrived, moved his family into, and began the restoration of, The Hall (the most distinguished building in the town); and started work immediately on converting the Kingston Mill for the manufacture of rubber goods. Built in 1811 from West Country stone and a full five storeys high, it was eminently suitable for its new role and was to remain the centre of Bradford's rubber industry for over one hundred years.

Stephen Moulton and his wife Elizabeth, looking over The Hall and the rubber works from the Tump on the edge of town. In the foreground are their daughters, Kate and Elizabeth. Painted by Elizabeth Tackle. (AMCT/DF)

3 Establishment and Expansion

"if you could see the number of rolls strewing our foundry yard you would not complain"

Bilston Foundry to Stephen Moulton, July 1849

It is to be noted that, unlike the weaving of cloth, rubber manufacture on a domestic scale is not a viable proposition; the equipment required being too costly and too large to contemplate anything other than production on an industrial scale. The expertise and experience required to equip the Kingston Mill were largely supplied by Moulton's friends and partners in America, the Rider brothers. In 1847 William Rider set out an "Estimate of Cost of a Rubber Factory" – this covers all the essential machinery and is worth reproducing in full here as an aid to understanding both the process and the associated costs: (the $:£ exchange rate being roughly 5:1)

Estimate cost here for a Rubber concern that will employ two hundred hands – say 160 girls and 40 men, who can turn out from ten to fifteen hundred Dollars worth of goods such as we have been making per day of 10 working hours.

1	Sixty horse power with boiler and setting	$6500
1	Spreading Machine	$3000
1	Softening Machine	$1500
2	Grinding Machine	$1000
1	Mixing Machine	$500
	Shafting, belting, pipes etc. for above	$1200
1	Cutting and Washing Machine, tube, belts etc.	$600
1	Cylinder Boiler Heater 3 ½ ft. diameter, 12 ft. long, with car for goods, valve etc., etc.	$1200
	Steam pipes for heating factory	$1500
	Fixtures for lighting factory	$200
	Tables, wash tubs, Tools of all kinds	$2500

		$19700

An establishment with the above machinery and hands would require to be equal to two hundred feet long forty feet wide and four stories high, and should have a glass platform, or room on the roof all equal to 150 feet sqr for sunning of goods – unless we get clear of that business – which I think we shall do – the establishment should have a good supply of water for washing rubber – In putting up factories, the expense always over runs estimates, then I think very probably the (average) fixings will cost a cool $25,000 here.

Rider's specification covered the equipment needed, and to aid the understanding of how the machinery would be used we can add some detail of the manufacturing process from raw rubber to finished product.

The Kingston Mill with The Hall behind. The River Avon is in the foreground. The unusual octagonal building shown is known today as 'The Dovecote' but was built as a wool drying stove. Painted by Elizabeth Tackle around 1850. (AMCT/DF)

Firstly, the raw material needed to be cleaned of impurities. The bulk rubber would be cut into small pieces by hand and the larger pieces of foreign matter removed – rubber was sold by weight and the addition of a few large stones would nicely bulk up a bundle to the advantage of both the harvester and the shipper. These cuts of rubber would then be fed into the water-filled 'Cutting and Washing machine' and as the knives tore the material into small fragments the debris – stones, sand etc. – being heavier, would fall to the bottom whilst the now clean rubber rose to the top.

Next came the 'Grinding machine' where the rubber was ground and compressed between two heated rollers. The rollers turned at unequal speeds to knead the rubber more effectively but this process was still very time-consuming, hence Rider's suggestion of two Grinding machines. The next two processes, Softening and Mixing, add the compounding agents to the rubber mix. In this way, the physical properties of the rubber compound can be adjusted to suit the end product. The most essential additive is of course sulphur, as this is required for vulcanization. Others include fillers – one of the most ubiquitous being the reinforcing agent 'Carbon Black'. Whilst the 'Grinding' and 'Softening' machines used steam-heated rollers to aid their functions, by necessity the 'Mixing' machine had water-cooled rollers to avoid partial vulcanization of the rubber compound.

The 'Spreading machine' specified by Rider is what we now know as a 'Calender'. This, the largest and most impressive item of processing plant and somewhat similar to an enormous clothes mangle, could be used to roll sheets of rubber to precise thicknesses; or to produce rubber-impregnated fabric – similar to McIntosh's oilskins – without the use of any solvents. The rollers on this sixteen-ton machine were made by the Highfield Foundry in Staffordshire and the frames were cast by Bush in Bristol. Other parts, including the wrought-iron tie rods, came from Coalbrookdale.

The 'Iron Duke' calender machine installed in the Kingston Mill (AMCT)

The final process was Vulcanization – the application of heat at a specific temperature for a measured amount of time to allow the sulphur to bind the compound into the stable cross-linked structure of vulcanized rubber. This could be done in a heated chamber (similar, other than in size, to a domestic oven) or, for large mechanical items, in heated moulds.

Moulton closely followed Rider's advice whilst fitting out the Kingston Mill; most of the designs and drawings for the plant came from his partners in New York who, despite their earlier reluctance, continued to support the 'English Project'. Indeed, Rider was so concerned that the machinery installed was of the best quality that he arranged for Mr William Frost – the man responsible for building the machines at the Riders' own factory – to travel to Bradford to supervise the design and commissioning of the new plant. Frost's employment terms were unusual in that Moulton offered him a quarter share of the profits in lieu of salary for his "time, trouble, skill and attention at the works", and allowed Frost to draw advance funds against these hoped-for gains. In 1850, Frost was joined by fellow Americans Miss Amelia Fisher and Mr S. P. Abbot, the latter being an experienced calender hand.

Rather than having complete machines built for him, Moulton opted to purchase parts from appropriate suppliers and have his own millwrights to erect them in the mill. There are many examples of terse exchanges between Moulton and the foundries contracted to cast and grind the rollers for his machines; the latter would complain of "an expensive experience" and "a very serious loss … if you could see the number of rolls strewing our foundry yard, you would not complain", whilst the former had no hesitation in suggesting to these companies how they should do their jobs and reminding them of consequences: "I am truly glad to hear that two of the rolls are finished – but how do you intend to grind the third? They must be ground together in the same frame, or otherwise they will not be true – any mistake now would be indeed most serious to me … as everything depends upon the truth of the faces of the rolls…" Despite these setbacks, the Kingston Mill was equipped for a sum considerably lower than that suggested by Rider and the machinery – and in particular the troublesome rolls – was judged to be significantly superior to that used by the Riders in New York.

Stephen Moulton and Co. of the Kingston India Rubber Mills opened for business on 3rd October 1848, with one of their first products being rubber-covered canvas belting which was very much in demand in mills, mines and machine shops. Staff numbered 23 of which 2 were female. It is likely that many of the men were millwrights and were installing and commissioning the machinery, as by early 1851 Moulton employed 40 men and 60 women. The output at the time consisted mainly of waterproof capes, jackets and other items of clothing; and from 1853 was to include thousands of capes, blankets and similar articles for the Crimean War.

Raw material, the latex tapped from rubber trees, had to be imported to England. In the early days, Moulton bought South American rubber (largely from *Hevea brasiliensis*) but before long he was also buying rubber from Africa. Despite the Bradford Works proximity to Bristol, little of the rubber trade came that way and Moulton predominantly bought from London and, later, from Liverpool. Due to the variability and inconsistency of raw rubber shipments Moulton was, wisely, wary of buying by description. He knew well enough that terms such as 'fair average, and merchantable condition' were as good as worthless and the only way to be certain of

S. MOULTON & CO's.

LIST OF PRICES

OF

INDIA RUBBER GOODS,

MANUFACTURED AT

THE KINGSTON MILLS,

BRADFORD, WILTS.

COATS.

Sizes assorted, viz.

No. 1. —	Chest	39 inches,	Waist	36 inches,	Length	40 inches.
2.	..	41	..	38	..	42
3.	..	43	..	40	..	44
4.	..	45	..	42	..	46

BLACK or DRAB, double and single breasted, each .... 22s.

CAPES,

Black or Drab.

No. 1.	Length 28 inches ..	.. 10s 0d	No. 4.	Length 40 inches ..	.. 16s 0d
2.	.. 32 ..	.. 11s 6d	5.	.. 45 ..	.. 19s 0d
3.	.. 36 ..	.. 13s 6d	6.	.. 50 ..	.. 23s 0d

Sleeves .... 2s 6d extra.

LEGGINGS.

Shirred at top	8s 0d
No. 1.	8s 0d
No. 2, with Belt	10s 6d

OVERALLS.

Nos. 1, 2 and 3, each 15s 0d

PANTALOONS.

Nos. 1, 2 and 3, each 14s 0d

Ladies' Capes, with Hoods.

Black or Drab.

No. 1.	Length 20 inches ..	.. 10s 0d	No. 4.	Length 32 inches ..	.. 13s 6d
2.	.. 24 ..	.. 11s 0d	5.	.. 36 ..	.. 15s 6d
3.	.. 28 ..	.. 12s 0d	6.	.. 40 ..	.. 18s 0d

Spanish Serappas or Ponchos.

Black or Drab	18s 0d
Ditto, with Hood	21s 0d

[OVER.

1192/570/1

Stephen Moulton & Co., Price List 1853 (WSA)

quality was to inspect and sample the material. Hence he eschewed the small savings that could be made by buying 'on shipment' and preferred to sample and buy on 'spot price' on landing (often attending the sales himself); therefore avoiding any concerns over quality and indeed actual delivery of the goods promised.

In early 1850, the Rider brothers were seeking to formalize the 'English Project' and William wrote to Moulton on this point: "The Kingston Mill must make some arrangement with us ... or it will never prosper – think of this and make up your mind that half or even a quarter of a loaf with security, is better than none..." Later in the year, a partnership agreement between Moulton and the three Riders was set up under which they would be joint traders and partners in the manufacture of rubber goods in Bradford on Avon. Moulton would provide the capital and would receive a fixed annual rental. Despite considerable discussion, Goodyear himself was not directly involved. This agreement covered a duration of four years and was not renewed (and, for reasons that will become apparent later, did not run to term); details are scarce as the original document is not traceable. It appears that the Riders would take 3/16ths of the profits in return for their (largely technical) involvement, and Frost was persuaded that a 3/16th share of a Moulton – Rider partnership was worth more to him than a quarter share in Moulton's company.

What is clear is that Moulton manufactured under his own English Patent using Hyposulphite of Lead, rather than licensing from Goodyear or Hancock. Emory Rider left America and came to live in Bradford to assist with Management and Technical Supervision at the Kingston Mill. His position clashed with that of Frost and, despite their previously amicable relationship in New York, the two men fell out. Frost was angered by Rider's status as partner and his encroachment on Frost's territory at the factory; Rider, in turn, was irritated by Frost's share of putative profits being three times that of his own and expressed dissatisfaction about the quality of Frost's work. This situation of simmering resentment persisted into 1851, further aggravated by Frost continuing to draw advances against his share of profits. Moulton eventually confronted Frost on the size of these withdrawals, having observed their magnitude with some alarm. Thus challenged, Frost's diligence at the Mill declined and in November of 1851 Moulton dismissed him without notice. His patience was exhausted and his hand was forced, for in that year alone Frost had drawn an advance of £591.13.2 against a total declared profit of £510.

In these early days, Moulton sold the bulk of his products (clothing) through general merchants who also had distribution networks in place. He desired, however, to reduce his dependence on these middlemen and was prepared to accept orders from any source – with preference to direct trade with large businesses such as railway companies but not to the exclusion of small orders from retail shops or even individuals. This made his sales administration – accounting, despatch, credit control – rather complex and his correspondence file somewhat lively. He later employed commission agents to visit industrial customers and demonstrate to them the advantages of using

Moulton rubber products; these travelling salesmen were paid 5-10% of the invoiced value of the goods.

To add another facet to his operations, in 1850 Moulton opened a retail salesroom in London at No. 2 St Dunstan's Terrace. This was a ground-breaking move and thought to be the first of its kind in England. As well as wishing to add the retail profit to that of manufacture, Moulton wanted to showcase his products – rubber was, of course, very novel and the general public were largely unaware of its uses and advantages over traditional materials, particularly with respect to clothing. Both McIntosh and the North British Rubber Co. later followed this early example of vertical integration and opened their own shops in the capital.

Business was brisk enough for Moulton to expand his manufacturing premises - from 1851 until 1860 they occupied the old woollen mill in Staverton, and in 1855 they took possession of the 'Middle Mill' in Bradford. Around this time, Moulton also acquired the Greenland Lower Mill in the town, which may have enabled him to vacate the Staverton property (he needed to, as we shall see later) and concentrate his manufacture in Bradford. This expansion paints a picture of a flourishing successful business - which it was, at least in a long-term strategic sense. Moulton was, however, beset by cashflow problems rather more frequently than was desirable. Much of this was due to the vagaries of the business viz. raw material supply and timely settlement of accounts by debtors. The larger merchants could be counted on to pay on time (indeed, Moulton was often able to persuade them to advance funds before manufacture, thus financing his work in progress) but the smaller ones, and the direct customers, were not so reliable.

S. MOULTON AND CO.,

Manufacturers of

CURED INDIA RUBBER BUFFERS, SPRINGS, STEAM PACKING, WASHERS,

HOSE TUBING, GARMENTS, BLANKETS,

ELASTIC HOT-WATER BEDS, CUSHIONS, &c., &c., &c.,

KINGSTON INDIA RUBBER MILLS,

BRADFORD, WILTS.

Stephen Moulton & Co., advert 1859 (MUS)

The Wilts. and Dorset Bank were accommodating to the tune of several thousand pounds, subject to their loans and overdrafts being secured on the Moulton's stock of crude rubber. In November 1853 the outstanding balance at the bank was £4,150 and by 1860 it was nudging £8,000 – more than Moulton had paid for the Kingston

Mills and The Hall. Given the forward payments from the merchants, and the secured loans from the bank, there were several times during the decade when every scrap of rubber in the raw state, all the work in progress and all the finished stock did not belong to Moulton at all. Indeed in 1853 the bank extended a further £1,000 to facilitate the continuing supply of raw rubber, subject to security being available in no more than ten weeks.

Aside from Palairet, Moulton's main backer was George Foster of G. H. Foster & Co., General Merchants of Moorgate Street, London. Foster advanced £11,000 in 1851, secured on the Kingston Estate and the new rubber manufacturing equipment. Foster had previously financed the manufacture of rubber goods for one of Moulton's London merchants and had sensed an opportunity; and indeed his loans were on a rather more commercial basis than Palairet's both in terms of repayment and security. Nevertheless, Foster had to reprimand Moulton for valuing his finished stocks (which were being offered as collateral) at selling price rather than at cost – "You cannot submit profits not realized." Following Foster's death in 1859, the balance outstanding – £6,000 – was transferred to the faithful Wilts. and Dorset Bank.

Moulton's false declaration of value of finished goods may have been either an oversight or an act of desperation, for the Kingston Mills made precious little actual profit in those early days – if indeed they made any at all. His business partners, the Rider brothers, became twitchy about their liabilities and wrote to Moulton in 1852 suggesting that the partnership agreement be terminated in advance of the four-year renewal date. Moulton, somewhat surprisingly, refused – he had not been keen to enter the partnership in the first instance and one may have thought he would have welcomed its dissolution. Perhaps, following the departure of Frost, he still needed Emory Rider at the factory to provide technical expertise as his (Moulton's) sons Alexander and Horatio were still learning the business.

Frost, meanwhile, was still furious over his dismissal. He claimed that he was a partner in Moulton's business and in 1852 sued for his share of the profits as agreed in 1848. The absence of his name in the articles of association did not help his case, and neither did his lack of appreciation of the actual profit levels – which had not escaped the noticed of Moulton's real partners, the Riders. Nevertheless Moulton generously offered to cover Frost's notional salary from the date of his leaving until he found another suitable job, together with a further £500 inducement – a sizeable fortune at the time - if he took a role in an unrelated industry. Whether he accepted or rejected this bribe is not recorded.

The munificent terms of Moulton's settlement with Frost may have influenced the Riders to take the legal route to rid themselves of the 'English Project' that they had so often cautioned Moulton about. Citing overspending, withdrawal of cash, failure to obtain capital funding and mounting losses, in 1853 the Riders pleaded with the court to dissolve the partnership on the grounds that Stephen Moulton & Co. was insolvent and thus any profit-sharing agreement was null and void. An example was made of the Staverton Mill, which Moulton had agreed to buy in 1851 but – beyond the deposit - was unable to obtain funds to complete the purchase.

The findings of the court are unknown, but Emory Rider returned to New York towards the end of 1853 and the partnership agreement was not renewed in 1854 as per its four year term. Moulton will have known that the Riders would not give their assent to renewal and he may well have reckoned that their job was done: his rubber factory was established and operational; and he possessed enough skill amongst his own staff – including his sons – to carry out his business without assistance from others. He was grateful for their effort and expertise in getting him into that position, but could now proceed without interference and with all the potential profits available to him to re-invest in the growth of the company.

4 Setback and Resurgence

"I know of none other than Moulton of Bradford on Avon, Wiltshire, who can make the right sort of material."

I K Brunel

Having developed, proved and patented his hyposulphite of lead vulcanization process and established his rubber factory in Bradford on Avon, Stephen Moulton may have been looking forward to being able to concentrate more on family life and the ongoing restoration of The Hall as well as the ongoing expansion of his business. He was to be disappointed, as the validity of his patent was soon to be questioned.

Moulton cannot have been completely surprised to find that Thomas Hancock was prepared to aggressively assert his patent rights as he had been warned as early as 1849 that Brockenden (Hancock's co-director) "intended to stop him as soon as he starts, and moreover he does not have a leg to stand on". Hancock's position was also set out unequivocally in his book of works in 1847: "This is where I stand. Certainly you may use my process to manufacture any of these (or related) products. Just remember to ask first and pay for a licence."

Moulton and Hancock faced each other in court on two occasions, both with slightly unusual results. In 1852 the Judge found 'very considerably' in favour of Hancock, but was not prepared to grant an injunction against the production at the Kingston Mills – "weighing all these considerations, and admitting the strength of the plaintiff's evidence he was of the opinion that he should not be warranted in committing the defendant without a trial at law on the question of infringement". In essence, he did not consider himself able to judge whether Moulton's process was "the same or substantially the same" as Hancock's. Moulton was left in no-mans land, being able to continue production but with the threat of legal action and the possibility of retrospective royalties and removal of any manufacturing rights hanging over him; patentees had a right to refuse to grant licences until 1883. Furthermore, any customers of the Moulton Rubber Co. were liable to be sued by Hancock for infringement. One cannot envy Stephen Moulton's position, or that of Palairet, Foster and the Wilts. and Dorset Bank. Finally, the rubber industry rose against Hancock and issued the writ of 'scire facias' – the most severe test of legality of a patent.

Thus in 1855 Hancock, Moulton and Goodyear stood in court; Hancock being uncomfortable in his unusual position of defendant. Other than for the Great Exhibition in 1851, Goodyear had been reluctant to travel to England before, so much so that Moulton had been moved to write to Rider in 1848 that "Goodyear is coming, of course so is the year 46732009, the latter will arrive first, I opine..." Goodyear made his case well, as this was his final chance to obtain to any rights to vulcanization in England, but it could not be proven that Hancock filed his patent before knowing the process. The jury found for Hancock and the protracted legal arguments were over. With Hancock's patent confirmed as valid and pre-dating both Goodyear and Moulton's patents, the latters' were now invalid and worthless. Again the court refused an injunction, preferring to insist that Moulton paid Hancock a licence fee for the (restricted) use of his (Hancock's) patent. This fee was set at £600 per annum – a substantial sum; Moulton's wage bill at the time was scarcely twice that. It does not appear that this fee was charged retrospectively. Despite this settlement, major changes were required at the Moulton Rubber Co. as this licence did not include vulcanized clothing or footwear, hitherto the mainstay of Kingston Mills manufacture. Moulton needed a new outlet for his rubber or he faced ruin. He already had a small business in rubber 'mechanicals' and he was forced to develop this much further – this change of direction was to define Bradford on Avon's output for the next one hundred years.

An interesting footnote to the patent position is that Goodyear patented in Scotland before Hancock did – and scarcely had the courts confirmed his patent position in England when the American ship Harmonia docked on the Clyde bringing the equipment and workers to inaugurate Scotland's rubber industry, using Goodyear's process. It is also notable that, following the full story of Hancock examining Goodyear's rubber samples before developing his patent being recounted in the courts, there were few interested parties in the rubber industry and the legal profession who did not take a dim view of Hancock's behaviour. Even whilst summing up the case, Lord Campbell suggested that "If Goodyear's invention was prior in point of time, it was not handsome in Hancock to look at his specimens and try to find out his discovery; and if Goodyear was the inventor it was to be regretted that he should not have the benefit of the invention." Goodyear was later to come to England, setting up a residence in Bath and an office in London. Moulton supplied many products in significant quantities to Goodyear over a period of two years, but eventually – like many before him – he became tired of Goodyear's ability (or lack of) to pay his debts.

Moulton's early business (before and during the patent wrangles) was largely textile-based, making weatherproof materials for clothing, footwear, sheeting and tarpaulins – and even water beds. Much of this material was manufactured by the immense 'Iron Duke' calender machine in the Kingston Mill. There was also a demand for rubber belts (which were significantly cheaper than the leather belts used hitherto), saddles and harness for horses, trolley tyres, cart springs, rubber buttons, dental vulcanite and rubber hoses which were used in many industries, particularly the brewing of beer. The nascent railway companies had used rubber as a spring medium for buffer springs even before vulcanization was discovered and it was not long before this industry

The Rubber Works - Kingston Mill with The Hall behind. A postcard by R. Wilkinson, showing clearly the typical arrangement of the Mill being overlooked by the owner's house. (WBR)

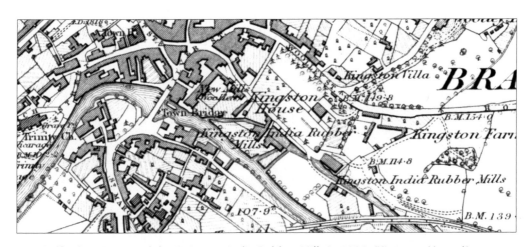

Bradford on Avon and the Kingston India Rubber Mills in 1886. 'Kingston House' is now called 'The Hall', 'Kingston Villa' is 'The Dower House'. The railway runs along the bottom of the map and the Lower Greenland Mill is just above it.
Whilst the Mill still stands, the railway was built over part of it and its mill race.

demanded more rubber products including hoses, rings, bearing springs, draught excluders, carriage interconnects and door stops. Moulton manufactured and sold many of these items in small quantities and began to establish a reputation amongst railway companies as a reliable supplier of consistently durable rubber products. This was to provide him a lifeline as he struggled to work with the terms of Hancock's limited licence.

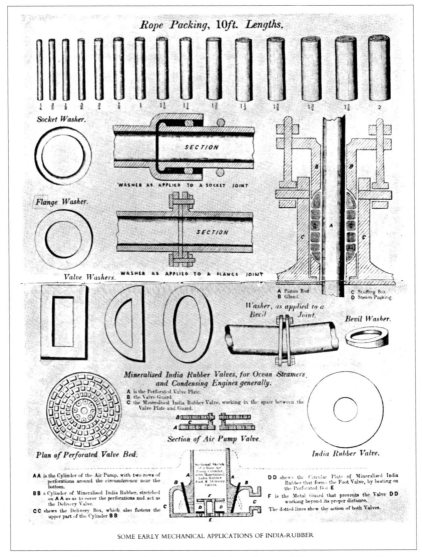

An illustration of early uses of vulcanized rubber components

An unusual request for Moulton's expertise in rubber mechanicals came from one of the most remarkable and prolific men in engineering history. Isambard Kingdom Brunel wrote to Stephen Moulton in 1859 seeking assistance with the manufacture of rubber mounts for the masts on his monumental steamship Great Eastern. He noted in his diary "I know of none other than Moulton of Bradford on Avon, Wiltshire, who can make the right sort of material". Later in the same year, Brunel endorsed Stephen Moulton's application for membership of the Institution of Civil Engineers and he was duly elected. Brunel, however, did not approve of the use of rubber in buffer springs and as a consequence the Great Western Railway did not specify them, despite most other railway companies fitting them as standard. This resistance continued through to Brunel's successor at the GWR, Daniel Gooch; both were, however, happy to use rubber for bearing springs (which was somewhat contrary to conventional thinking).

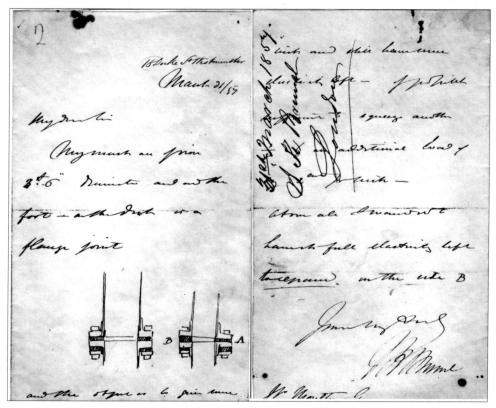

Letter from Isambard Kingdom Brunel to Stephen Moulton ref:
Great Eastern steamship,1859 (AMCT)

Turning back time a little, it is notable that Goodyear, Hancock and Moulton all displayed their wares at the Great Exhibition in 1851 – the former with characteristic exuberance and flamboyance, despite his straightened circumstances. The showcase of this celebration of industrial technology and design – the Crystal Palace – was designed by Sir Joseph Paxton and the contractors were Fox, Henderson & Co. The lead partner at Fox, Henderson was the eminent railway engineer Sir Charles Fox (who, amongst other innovations, conceived the first railway points) and his understudy was George Spencer. Spencer left Fox, Henderson in 1852 and established George Spencer and Co. as a designer and supplier of mechanical parts for the now burgeoning railway industry. Spencer saw a great quality in rubber – it could absorb almost any blow without breaking or permanently deforming – and his first patent was for a rubber buffer spring. The design of this spring – a double cone shape constrained by a cast iron annular ring – confirms that, even at this early date, Spencer had a deep understanding of the properties of rubber as a springing medium and how best to utilise them.

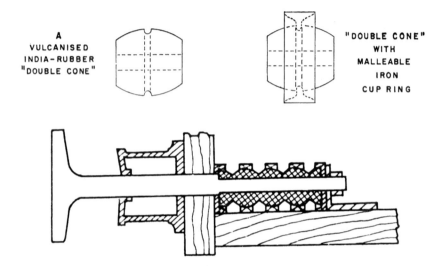

Geo. Spencer & Co. double-cone rubber spring and an example of how a buffer would be assembled. Increases in train loads drove further innovation in later years.

It will be appreciated that vulcanized rubber springs had many advantages over the traditional steel helical (coil) spring – as well as the enormous deformation permitted in the material, rubber springs were not liable to fracture under heavy impact loads. They required no lubrication and did not corrode, were lighter, and were excellent at absorbing and damping shocks. Furthermore, they were considerably cheaper than steel springs, at least until the commercialization of the Bessemer process for the production of steel in the late 1850s.

Spencer did not manufacture himself, choosing to sub-contract the work to existing rubber manufactories. He did his research into suppliers carefully and placed his business with the longest-established and most reputed company – McIntosh in Manchester. In common with his customers, he became increasingly frustrated by the variable quality of these early rubber products and sought consistency of quality over price and availability – engineers in railway companies were not disposed to specify components where performance could not be predicted. The situation came to a head in 1857 when the London and North Western Railway returned a batch of springs with the terse report that they had "no more elasticity than a block of wood". Such defects were clearly the fault of the rubber manufacturer in failing to maintain quality of raw materials and control the vulcanization process, rather than the intrinsic design of the rubber springs themselves, yet it was the reputation of Spencer that was at risk from such failures. Spencer's agents found that Moulton's rubber mechanicals supplied to railway companies were held in high regard and often specified in preference to their own for reasons of reliability and durability.

George Spencer placed his first sample orders with Moulton & Co. in July 1858. There will have been a reluctance to do so – as well as the considerable cost of providing new moulds, a degree of courage was required to break with the most eminent rubber manufacturers in the country. This hesitation is confirmed by Spencer supplying some of Moulton's moulded springs (and hyposulphite of lead, the vulcanizing agent used by Moulton in his own – now invalid – patent) to McIntosh for chemical analysis and mechanical tests in an attempt to discover the reason for the superiority of the Bradford on Avon product. This was to no avail. Moreover, the springs made by Moulton were so well received by railway engineers that Spencer was moved to place the bulk of his business with his new supplier.

George Spencer
1810 - 1889 (DTR)

Following the loss of their exclusivity on the Spencer spring contract, McIntosh & Co, dealing direct with railway companies, aggressively pushed the sale of their own simple buffer spring (which was of rather less technical worth than Spencer's double cone). In a further act of treachery, they did this by citing the high failure rate of the Spencer product – which of course was due to their own failures in manufacture, and indeed their own design springs suffered from the same faults. George Fenelle – who led Spencer's marketing team – was moved to write to his boss in unusually blunt language "I could not conceive that such unprincipled practices would have been put in action … they not only do us great injury by their neglect in the case of those cones but take advantage to themselves for their own wrong and endeavour to injure us further … this is so disgusting that I cannot help thinking the sooner we have done with them the better."

Despite this, there was to be no 'clean break' as the demand for Spencer's patented buffer spring was so great that at times every available mould at Moultons and McIntoshs was pressed into service. However, McIntosh merely made what Moulton could not produce in a timely fashion. Following the introduction of George Spencer's improved cylindrical spring in 1866 new moulds were laid down, and all of these were at the Kingston Mills in Bradford on Avon. From this point onwards, Moulton supplied almost all of Spencer's requirements except for some specialist articles from other rubber manufacturers. These items dwindled to tiny numbers and by 1877 Moulton had agreed to act as factors in purchasing these for Spencer. The accounts for 1876 show that McIntosh's sales to Spencer were only £22, whereas Moulton's Spencer account totalled £27,515.

GEORGE SPENCER & CO.,
77, CANNON STREET, LONDON, E.C.
Mechanical Engineers
FOR INDIA RUBBER APPLICATIONS,
MANUFACTURERS OF THE
PATENT CYLINDRICAL INDIA RUBBER BUFFER & DRAW SPRINGS.

FIBROUS STEAM PACKING, HOSE PIPES,
AND ALL KINDS OF
INDIA RUBBER ARTICLES
FOR RAILWAY AND OTHER MECHANICAL PURPOSES.

SOLE LICENCEES FOR
G. ATTOCK'S PATENT CARRIAGE BODY SPRINGS, AND FOR WM. ADAMS'S PATENT BOGIE FOR ENGINES AND CARRIAGES.

George Spencer & Co. advertisement. As well as 'Mechanical Engineers for India Rubber Applications', they claim to be 'Manufacturers of the patent cylindrical india rubber buffer and draw springs' – all of which were made by S. Moulton & Co. and Chas. McIntosh & Co.

George Spencer's understudies were his heirs – his sons Alfred (George) and Alexander were most active in the technical department whilst his third son, Frank, took a role in sales (he would later become Sales Director). Alfred was to become a partner in the business in 1877, followed by Alexander in 1882 and Frank in 1889. The Spencer Company continued to innovate and new designs for buffer springs – introduced roughly every decade – were gladly adopted by railway companies as train loads rose inexorably. George Spencer's embedded plate spring of 1877 was acknowledged as being far more efficient than any previous rubber spring; Alfred Spencer's 1886 compound spring improved on this further by eliminating all metal-to-metal contact.

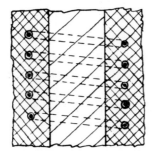

Stephen Moulton's patented composite embedded coil spring - drawing and photograph of a surviving sample. (DF)

As all of these designs were patented, Spencer had tight control of the market and - of course - all were manufactured at the Kingston Mills.

It is of note that, in addition to the principal business in rubber springs for railway applications, Spencers established themselves as 'rubber engineers' and suppliers of a miscellany of valves, belts, washers, hose etc. George Spencer foresaw a great potential in the application of rubber in rolling stock construction and moved to supply this market with his own creations or by obtaining the rights to the designs of others. Thus in 1863 Spencer & Co became the sole suppliers of George Attock's body cushion springs ('Attock's Blocks') that were widely used between the underframes and bodies of railway wagons; in 1864 the draught excluder patented by John Wilson became a Spencer product. Alfred Spencer designed and patented 'anti-rattle' window seals and trolley wheels. Alexander Spencer's early successes included the conical door stop (the like of which can still be seen on heritage trains) and high-pressure reinforced hose for vacuum brake systems (both from 1886).

Through their network of agents under George Fenelle, the Spencers maintained close links with the railway companies and their engineers. As well as keeping the company abreast of any technical developments – and hence often early options on new designs – this gave Fenelle many opportunities to foster goodwill with his customers. As well as recommending the inventions of others to railway officials, he would often pressure the Spencer Co. to show an interest in (or indeed to manufacture) the designs of influential railway engineers - the primary objective often being not the sales of the items themselves but the fostering of goodwill and the expectation of reciprocity. Whilst the firm showed little interest in designs by inventors from other fields, railway engineers with power and influence were assured of substantial co-operation. Thus Spencer obtained the rights to many patents – of greater or lesser worth - on the basis that "if it leads to his adopting our cones, it will be of some benefit".

As well as supplying most of Spencer's rubber mouldings, Moulton also manufactured other designs, including his own. In 1861 he patented a composite spring consisting of a helical steel spring embedded in a cylinder of rubber. Whilst this had some advantages over other designs, it was not economical in manufacture and never gained wide acceptance in the marketplace – somewhat due to Moulton's flourishing alliance with Spencer. Other Moulton patents include a device to lessen the recoil of a cannon by the use of rubber springs (1863) and a patent for sponge rubber in 1868.

The degree of collaboration between Spencer and Moulton can be ably demonstrated by numbers: in 1866 Spencers were purchasing 80% of their products from Moulton; representing 35% of the latter's total sales. Moving forward to 1876 the numbers were 95% and 80% respectively and would increase further. Clearly both companies had become interdependent although, to a degree, Spencer held the upper hand. It is to be noted that they appear not to have abused this position to any significant extent (although price reductions were often requested and usually granted), but naturally Moulton was wary of so much of his business being through one customer and the risks that it entailed. After all, Spencer could place business elsewhere without too much difficulty, albeit with some disruption costs.

Chief amongst reasons why Spencer did not seek other suppliers - and there were many who courted them - was Moulton's unmatched quality and obliging attitude. Still haunted by the early failures of buffer springs manufactured by McIntosh, Spencer would replace any defective products free of charge and would expect their suppliers to do likewise. They would expect rubber springs to have a working life of at least ten years. On one occasion a batch deteriorated after six years and Moulton wrote that "We are afraid that we cannot make them last forever; we will, of course, replace them if you think necessary" but did point out that "the failure of springs is not always our fault, as in the ordinary course of events they must be subjected to unfair usage at times."

It appears that both companies found it agreeable and profitable to work together, with Spencer concentrating on design and sales; and Moulton concentrating on manufacturing. Despite this, the position was far from ideal. Moulton was reluctant to invest in new equipment, and Spencer was frustrated by delays in production due to lack of production capacity.

The Kingston Mills India Rubber Works, 1899.
The buildings in the foreground were the first ones purpose-built for rubber manufacture.
Kingston House and the New Mills can be seen in the background. (DL)

5   Mixing and Rolling

"Well, Pickles, what is the price of Rubber today?"

John Moulton, Snr

George Spencer and Stephen Moulton first discussed the possibility of a merger of the two companies as early as 1879, but following Stephen Moulton's death in 1880, no further negotiations took place for several years. Alexander Moulton had been manager of the Kingston Mill operations since 1861 and remained in this position with assistance from his younger brother Horatio. Continuity seemed assured. Whilst business with railway companies in the United Kingdom had seen a measured decline as most railways were fully equipped with rolling stock, this was more than offset by new markets – notably India – and turnover rose sharply in the early 1880s. Spencer – and hence Moulton – were truly the world leaders in the supply of rubber mechanicals for the railway industry.

This apparent calm was disrupted in 1885 on the death of Alexander Moulton. A man of outstanding business ability, Alexander had worked with his father from the beginning of the English Project and his passing was a great loss to the company. Furthermore, Horatio Moulton expressed a desire to be relieved from constant attendance at Kingston Mill. Negotiations between the two companies re-commenced in 1887, but both parties agreed that it would be more tactful to wait for the passing of Stephen Moulton's widow, Elizabeth. In fact, further progress was not made until after the deaths of Elizabeth Moulton in 1888 and George Spencer in 1889. Three options were considered:

1.   A partnership between the two firms, working as one firm entirely.
2.   A partnership between George Spencer & Co. and Stephen Moulton & Co. at Bradford, but Spencer and Co. working separately at 77 Cannon Street, London.
3.   George Spencer and Co. to buy Stephen Moulton & Co. outright or by yearly instalments; the two Moultons, Horatio and John, retaining a share of profits for some years as sleeping partners, but Horatio remaining as manager for a certain period.

Spencers had for some time been troubled by their position as intermediaries in the supply of rubber items, and in some cases had declared themselves as partners in manufacture in an attempt to prevent their customers from dealing direct with Moulton. They were even moved to misrepresent the actuality in writing: "We manufacture all the spring rubber we supply at the works known for many years at Bradford on Avon, Wiltshire, as Stephen Moulton & Co., in which firm we are partners." This came to a head in a legal case in 1887, where Spencers realized that a disclosure of the truth would be likely to damage their reputation. To this end, Horatio Moulton was put under pressure to accept them as partners to strengthen their case. Despite their pleading, Horatio was unable to help them. Spencer wrote back "we very much regret this as we wished, when placed on oath, to be in a position to say "We are the manufacturers" ... should the question arise it is most likely to damage us considerably and, of course, you at the same time."

By 1890, negotiations were at an advanced stage and both parties preferred to move towards amalgamation. Spencers made their intentions clear as to why they were seeking a merger and John Moulton noted them down to discuss with his elder brother Horatio:

a) The natural desire on the part of Spencers for the purpose of their business to say with truth that they are Partners and therefore manufacturers of the articles they sell;

b) To be assured against any sudden interruption in the supply of the manufactured article on the death or resignation of Horatio Moulton;

c) So that they could become acquainted with the manufacturing process.

It was agreed that a new company would be created, incorporating all the assets of both existing companies, and that the shares would be apportioned according to the profits of the respective businesses over the last five years. On 22nd August 1891 the Articles of Association were signed and George Spencer, Moulton & Co. Ltd. was formed. Horatio and John Moulton were the largest shareholders, with 4,280 and 4,281 shares respectively; but the Spencer interests held the balance of 14,752 shares and hence 63% of the shareholding. The nominal capital of £233,310 (in 2015 – on a relative GDP basis, c. £300 million) did not include the land or buildings; these – and The Hall – remained the property of the Moulton family.

Like its forebears, the new company was very much a family affair. The Board of Directors consisted of Alfred, Alexander and Frank Spencer (all sons of George Spencer), Edward Stidolph (George Spencer's son-in-law) and Horatio and John Moulton (Stephen Moulton's sons). The latter two were Managing Director at Bradford and Chairman, respectively. George Spencer's youngest son, Sydney, was to join the Board at a later date.

Aside from the developments in Bradford on Avon, there were other changes in the rubber industry. Close to home, and no doubt encouraged by the prosperity of Moulton & Co. (and perhaps by their rather over-full order book), Giles and Willie Holbrow established a small rubber factory at Avon Mill in Limpley Stoke in 1875. On Giles' retirement in 1886 this small concern passed into the hands of Messrs. Browne and Margetson and began a rapid programme of expansion. By 1889 they had moved to much larger premises in Melksham and, as the publicly listed 'Avon India Rubber Co.', were producing solid tyres, rubber belting and railway mechanicals.

Railway mechanicals formed the bulk of the Kingston Mills' production for over a century.

George Spencer, Moulton & Co. (formed by the merger of the two companies) could rightly claim that they were both Mechanical Engineers and Manufacturers.
This advert is from 1900.

Furthermore, the demand for rubber was rising at a rapid rate. The arrival of two inventions combined to bring personal transport to the people in a similar fashion to how railways had brought public transport to them a generation before. First came the safety bicycle, designed by John Kemp Starley of Coventry in 1885. His 'Rover Safety' bicycle was a revolutionary step that moved cycling from the dangerous pursuit of riding a 'Penny-Farthing' – a preserve of the fearless and athletic – to a universal mode of transport. By the early 1890s, the Penny-Farthing was obsolete. Following hard on the Rover Safety's (rubber-tyred) tracks came John Boyd Dunlop's pneumatic tyre. Dunlop was a Scottish vet with a large business in Belfast, Ireland. He had used rubber tubing in his veterinary practice and, in 1887, experimented by adding tubes filled with pressurized air to the wheels of his son's tricycle.

Dunlop patented his invention and within a year Willie Hume, captain of the Belfast Cruisers Cycling Club, entered four cycle races on Dunlop's tyres and won them all.

In the crowd, Irish financier and president of the Irish Cyclists' Association Harvey du Cros watched his sons emphatically beaten by Hume in one of the races and knew an opportunity when he saw one. Du Cros and Dunlop formed a company which was eventually floated as the Dunlop Pneumatic Tyre Co. in 1896. Dunlop's patents were set aside in 1892 (Thomson had patented a similar invention in 1845) but the company's rise was spectacular – by 1896 Dunlop's nominal capital was double that of Spencer, Moulton & Co.

The new vogue for cycling brought a huge new demand for rubber which the manufacturers and raw material suppliers struggled to meet. The India Rubber Journal could not quite believe it: "There is certain to be some new invention which will arise and take the place of the pneumatic tyre at no very distant date when the vast amounts invested in the pneumatic tyre business will be practically sunk, if not altogether lost ... this cycling craze is only a nine days' wonder ..." How wrong they were. By the turn of the century, rubber tyre production accounted for one third of the entire rubber industry – and those waiting for the replacement for pneumatic tyres 'at no very distant date' would still be waiting today.

It is to be regretted that Spencer, Moulton & Co. were not well placed to take advantage of this new business. Spencer's history was in railway mechanicals, and the principal rubber 'brains' at the Kingston Mill in Bradford were all dead – Horatio Moulton having passed away in 1893, eight years after his brother Alexander and scarcely thirteen years after his father. John Moulton – Stephen's youngest son – kept an interest and a substantial shareholding, but he was a barrister-at-law, not a rubber factory manager. Notwithstanding his position as non-executive Chairman, it was said that his active involvement was limited to infrequent visits to the works (which were, of course, at the bottom of his garden) to enquire – "Well, Pickles, what is the price of rubber today?" Samuel Shrowder Pickles was Spencer Moulton's chief chemist from 1912 until 1950, and the man who discovered the theory – the sulphur cross-linking the long chain rubber molecules, and thus giving a permanence to the physical and mechanical properties of the material – behind the practice of vulcanization.

All four of George Spencer's sons continued to work within the firm, and, as was their intent when the merger talks were taking place, certainly became more involved with the business of rubber manufacture. Alfred Spencer became Managing Director following the death of Horatio Moulton. Sydney Spencer was appointed Works Manager around the turn of the century.

As the weaving trade continued its decline in Bradford, more mills were abandoned by the clothiers. Ward and Taylor vacated the 'New Mills' on the western side of the Kingston Mills site in 1898. Spencer Moulton purchased this building a year later and, in 1900, the weaving shed that adjoined it. The latter was converted to two storeys, the ground floor having stone-arched vaults which were used as a fire-proof store for the raw rubber. Fire was an ever-present danger – rubber itself is flammable and sulphur presents an explosion risk, and the fumes from both are unpleasant and

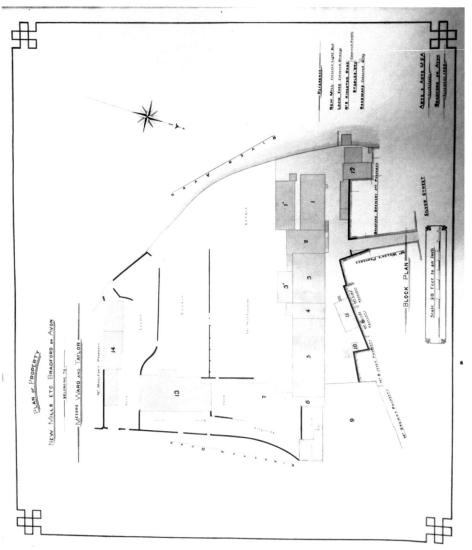

The New Mills Factory - Ward & Taylor Clothiers - in 1889. The Lamb Inn, owned by the Bradford Brewery Co., is shown bottom right. The Kingston Mills are off the top of this plan ('Mr Moulton's Property') and the Kingston Road is on the left. (WSA)

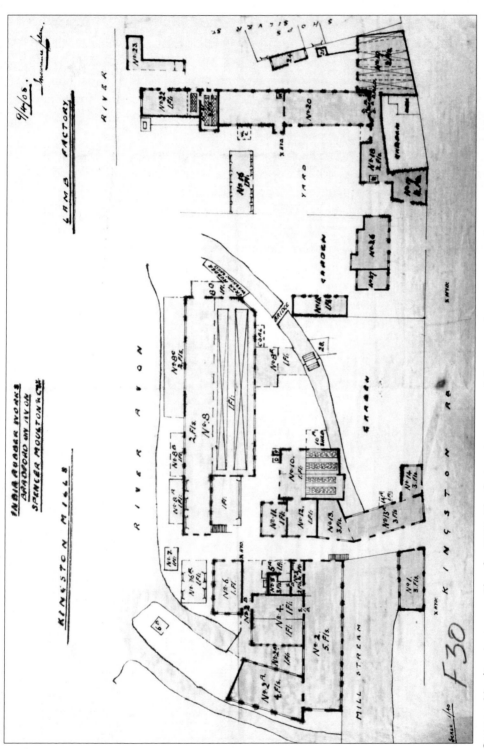

Spencer Moulton Insurance Plan for 1908. The factory is now much larger and encompasses the New Mills and The Vaults (shown here together as the 'Lamb Factory') as well as the Kingston Mills. The Kingston Mill itself is the 'No.2' building on the left. The River Avon is to the top, the Kingston Road at the bottom. The Town Bridge, and the town centre, are not shown but are off the plan to the right. (WSA)

hazardous. Spencer Moulton employed their own 'fire brigade' to deal with any incendiary incidents quickly and effectively.

The dirty business of mixing took place in the New Mills and the whole building (including the vaults) became known colloquially as the 'Black Hole'. This 'new' group of Spencer Moulton buildings at the western end of the site was labelled as the 'Lamb Factory' (named after the Lamb Inn by the bridge, and notionally distinct from the Kingston Mills) as early as 1908. Later plans would describe the whole factory east of the town bridge as 'Kingston Mills' as per earlier practice.

Working in the 'Black Hole' did have at least one advantage – in order to keep the raw rubber pliable the building was heated to a comfortable temperature. Conversely, of course, it was hot in the summer - not that this was unique to the 'Black Hole'. The heat from the moulding presses pushed the air temperature upwards in all the moulding shops and workers were paid a 'heat allowance' if the mercury rose above a set level. On days where it rose near to this it could be difficult to resist giving it a helping hand by placing the thermometer on the top of a moulding press for a couple of minutes shortly before the inspector came round. Another use of the top of a moulding press was as a hotplate; hunger pangs on a night shift could be satisfied by a bubbling can of beans, although they did take a little longer to heat up than on a more conventional cooker.

Rubber Compounding –
Weighing Carbon Black additive (DL)

44

Spencer, Moulton & Co. did enter the pneumatic tyre business (they had been making solid tyres for many years, however demand for these was later to be overwhelmed by the pneumatics) but very much 'stuck to their knitting' and kept focussed on their core activities. By contrast their neighbours in Melksham, the Avon India Rubber Co., grasped this new opportunity with both hands and began to produce pneumatic tyres for bicycles in 1900 – swiftly followed in 1903 by tyres for motor-cars. The automotive industry was very much at the embryonic stage and Avon Rubber grew rapidly with it, establishing itself as a leading and highly-respected tyre manufacturer.

Even earlier to the motor-car tyre business was the Sirdar Rubber Company, founded in 1899 at Baker Street, London with its factory nearby at Shirland Mews, Paddington. Sirdar did not make bicycle tyres, choosing to supply the larger sizes needed for carriages, charabancs and motor-cars and proudly listing 'HM the King, HRH the Prince of Wales and HRH The Duke of Connaught' amongst their customers. Initially these were solid tyres, as pneumatic technology had not yet developed enough to be safely used for tyres on heavy vehicles, but gradually load limits were raised as carcass design and moulding techniques – as well as rubber compounds – improved. In order to meet increasing demand, Sirdar opened a new rubber manufactory at the Greenland Mills in Bradford on Avon.

Greenland Mills lay to the east of the town centre, and both 'Upper' and 'Middle' mills had been occupied by the woollen cloth weavers J. W. Applegate (the 'Lower' mill being part of Moulton's factory). After the closure of Applegates – Bradford's last clothiers – in 1905, the mills lay empty until Sirdar arrived in the following year. It is testament to the reputation of Moulton at Kingston Mills that Sirdar chose Bradford as the location for its expansion – it was now very much a 'Rubber Town'.

Royal Sirdar Rubber Factory, Greenland Mills. The Middle Mill is in the
foreground and the Upper Mill is to the left. (AMCT/DF)

OUR NO. 3 LETTER TO THE TRADE.

CONTRACTORS TO HIS MAJESTY'S GOVERNMENT AND RAILWAY COMPANIES, &c

Telephones 6. BRADFORD on AVON
Telegrams "SIRDAR"
Code used A B C 5TH EDITION

HEAD OFFICE.
GREENLAND MILLS

MILLS-GREENLAND MILLS, BRADFORD ON AVON

DEPOTS
21. CRAWFORD STREET, LONDON. W.
GREAT SHIP STREET, DUBLIN
84. SAUCHIEHALL S⁺ GLASGOW
&c

Contractors to—
THE WAR OFFICE.
THE POST OFFICE.
THE INDIA OFFICE.
THE CROWN AGENTS.
COLONIAL GOVERNMENTS.
FOREIGN GOVERNMENTS.
L & N.W.R.
G W.R
M R.
L & S.W R.
G.E.R.
G C.R.
L C.C.
Etc.

Manufacturers of—
CARRIAGE TYRES.
MOTOR TYRES.
BICYCLE TYRES.
VALVES.
WASHERS.
RINGS.
TUBING.
SHEETING.
DOOR STOPS.
WINDOW STOPS.
BUFFERS.
Etc.

BRADFORD on AVON.   October, 1914.

Dear Sir,

Since addressing our letters No. 1 and No. 2 to the Trade, the War has brought about a new condition of affairs, and Solid Tyres of all sorts are coming rapidly into favour. Therefore the object of this our No. 3 letter is to bring directly to your notice, Royal SIRDAR Solid Rubber Tyres. as follows:-

SOLID MOTOR TYRES.

''Clinched-on'' or ''Royal Sirdar Buffer'' styles. Strongly recommended for pleasure cars and light delivery vans.

''Steel-Banded'' Tyres. These are supplied for light vans if required, but are more particularly intended for heavy commercial vehicles.

SOLID CARRIAGE TYRES.

We are now quoting very special prices for coils or short lengths of our well-known ''ROYAL BRAND'' Rubber, or if you do not fit your own tyres, we undertake this work and pay carriage on wheels one way.

It will pay you to write for prices and full particulars of the above. We will give your enquiries prompt and careful attention, and quick deliveries are guaranteed.

Yours faithfully,

THE SIRDAR RUBBER CO., Ltd.,

I. F. M.

Commercial Manager.

Sirdar letter, 1914

Royal Sirdar advertisement 1902

In 1911 the War Office awarded Sirdar a 12 month contract for the supply of pneumatic motor car tyres and inner tubes. There is some evidence to show that Sirdar used an enterprising business model – rather than selling tyres as an outright purchase, they would offer them on a 'cost per mile' basis. This guaranteed their durability and at least some customers found them to be cheaper than their rivals. Sirdar would fit the tyres to the customers' wheels themselves, to ensure correct fitment and minimise damage. Furthermore, if the tyres were fitted to a bus with advertising panels, the operator was obliged to add "This bus is fitted with Royal Sirdar Buffer Tyres". Sadly this progressive thinking did not save Sirdar from bankruptcy in 1914 and the Greenland Mills lay abandoned for a year before the Avon India Rubber Co. – and 300 staff – took over. This new venture was under the auspices of the Ministry of Munitions; so great was the wartime demand for rubber tyres that the factory workers were in the direct employment of the Government.

The Avon India Rubber Co. remained at the Greenland Mills until 1933 when they closed their Bradford operation and retreated to their main factory in Melksham. The mills were then occupied by Dotesios Printers, but that was not the end of rubber production on this site, as the Rex Rubber Company were in residence from 1947 until the late 1980s. Amongst other small items, Rex manufactured tennis balls. The balls themselves were moulded at the factory and a substantial number of local 'home workers' were employed to stick on the fabric covers.

6 Honour Vindicated

Following the declaration of war in 1914, John Moulton – youngest son of Stephen Moulton, and Chairman of Spencer, Moulton & Co. – presided over a recruiting meeting at the Town Hall in Bradford. Out of the 1,000 present, more than 200 men volunteered, with 120 of them being employed at the Kingston Mills. Those who signed up were given £10 – £5 from Spencer Moulton and £5 from John Moulton himself. By September men were boarding trains and being sent to training camps and to the front.

Position gave no exemption - amongst those who signed up was (Charles) Eric Moulton, younger son of John Moulton and Assistant Works Manager at Spencer Moulton, who was commissioned as second Lieutenant in the 6th Wiltshire Regiment. He was promoted to Lieutenant in July 1915. Staff from the Head Office in London who enlisted included George Spencer (only son of Alexander Spencer and already a member of the Board of Directors) and Richard Kenneth Glascodine. These two young men from the Technical Department represented the continuity of George and Alexander Spencer's innovative work in the design of rubber springs. Lieutenant Spencer joined the Artists Rifles in 1914 and, as a member of the Rifle Brigade of the London Regiment, was posted to France in October 1916.

Those who remained in Bradford had no shortage of work. Whilst there was a lull in demand for railway mechanicals, there was a boom in requirements for solid and pneumatic tyres. This was met both by Spencer Moulton and, at Greenland Mills, the Avon India Rubber Co. The Abbey Mill in Bradford, used from about 1898 to 1911 by a rug making firm was taken over by Spencer Moulton from 1915 onwards. In 1916, with space still at a premium, Spencer Moulton purchased the Lamb Inn and surrounding buildings by the town bridge. Having demolished these, a new factory – the 'Lamb Building' – was erected using prefabricated reinforced concrete sections, one of the first in the country. Bradford's mills were now almost totally devoted to the manufacture of rubber products.

As the war dragged on, more young men were called up to serve their country. Many never returned. Spencer Moulton lost 47 of its workforce, including Eric Moulton who was killed in action at Festubert in September 1915. As well as being a tragedy for his family, this was a bitter blow for the company as Eric was being trained to assume the mantle of Managing Director. Further great misfortune came in December 1917 when George Spencer died from wounds sustained in battle. All are remembered on a memorial plaque that is still affixed to the outside of Kingston House.

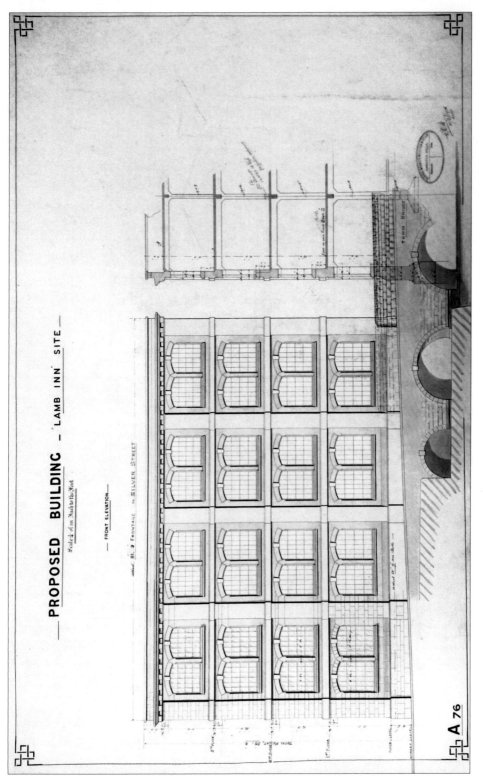

Plans for the new Lamb Building, four storeys. The eventual building had two storeys only and was one of the first in the country to be built using reinforced concrete framing, cast in London and erected on site. (WSA)

The "One Way"
to avoid "Hold-ups"

FIT
SPENCER
MOULTON

SPENCER
MOULTON
DUMB-BELL
CORDS TYRES
GUARANTEED
FOR
8,000 MILES
AGAINST
ANYTHING.

Geo. Spencer Moulton & Co., Ltd.,
2, Central Buildings, Westminster, S.W.1

Dumb-bells
give Grip
& Strength.

If your local Dealer
does not stock them
write direct for list.

The Perfect Non-Skids in Practice.

Spencer Moulton did not enter the rubber tyre business with anything like
as much vigour as their neighbours, the Avon India Rubber Co.
They did, however, offer a very bold guarantee. (AMCT/DF)

An unusual diversion occurred in 1916 when Spencer Moulton were accused of treachery by the esteemed organ the *Daily Mail*. It was alleged that Spencer Moulton were supplying tyres to neutral countries without adequate safeguards against onward shipping, and that these tyres were being used to aid the German war effort. The directors were incensed and took legal action against Associated Newspapers Ltd, the result being the court finding the allegations entirely without foundation. Spencer Moulton were further moved to place 'Honour Vindicated' advertisements detailing the story and adding that, far from supplying tyres without ensuring they did not fall into enemy hands, at the outbreak of war they had recalled their stocks of tyres from neutral countries.

April 12, 1916   THE MOTOR TRADER   85.

HONOUR VINDICATED.

In the course of the proceedings in the action for libel recently brought by George Spencer Moulton & Co., Ltd., against the Associated Newspapers Limited, as the Proprietors, printers, and publishers of the *Daily Mail*, the suggestions implying unpatriotic action on the part of George Spencer Moulton & Co., Ltd., and allegations that by some means or another they were supplying Spencer Moulton tyres to neutral countries without sufficient safeguards to ensure that such tyres should not be passed on for use on German cars, were unreservedly withdrawn, a complete apology was in the most honourable manner tendered and substantial damages and costs were agreed to. The honour and patriotism of George Spencer Moulton & Co., Ltd., were thus fully vindicated, but the Company desires to emphasize as strongly as possible the great point established by the evidence, namely, that when war broke out George Spencer Moulton & Co., Ltd., realising that Germany would be sorely needing tyres and rubber goods, immediately decided to stop the exportation of their tyres to any neutral country from which the goods might be transferred to the enemy. George Spencer Moulton & Co., Ltd., went even further than this, by endeavouring to get back their tyre stocks from their agents in Norway, Denmark, etc. In this effort the Company were unsuccessful as the agents paid for whatever stock they had on hand, and George Spencer Moulton & Co., Ltd., were thwarted in their attempt to " make assurance doubly sure " in the matter of preventing the leakage of their goods to an enemy destination. That a determined and patriotic effort was made in the national interest is, however, established, and George Spencer Moulton & Co., Ltd., are anxious that their friends in the British motor trade should have this proof of their *honâ-fides* brought clearly and unequivocally home to them.

GEORGE SPENCER, MOULTON & CO., LTD.,
77, Cannon Street, LONDON, E.C.

GLASGOW : 65-67, Bothwell St. LEEDS: 68, Albion St. MANCHESTER DISTRICT: Leo Swain & Co., Managing Agents, 287-289, Deansgate. NOTTINGHAM; A. R. Atkey & Co., Managing Agents, Trent St. IRISH DEPOTS: O. E. Jacob, Managing Agent, Dublin, Belfast, and Cork. Works; Kingston Mills, Bradford-on-Avon, Wilts.

Honour Vindicated' - Spencer Moulton's riposte to the *Daily Mail*

The cessation of war and return of peace saw railway companies licking their wounds and beginning to tackle the backlog of maintenance that had been accumulating since 1914. Demand for buffer and bearing springs soared and, with (in the UK) only the 'Big Four' railway companies in existence after grouping in 1923, administration became somewhat simpler as contracts became fewer – albeit with the importance of each contract increasing.

With quality, price and consistency of supply of raw material always being a primary concern, Spencer Moulton acquired their own rubber plantation - the Pundut Estates in Malaya. This covered a total of 2,860 acres and was seen not as a profit-making entity but rather as essential to lessen the company's exposure to the vagaries of raw material quality, price and delivery. Whilst Spencer Moulton still had to buy from other suppliers, the Pundut Estates were able to supply up to 60% of their requirements.

Mindful of how George Spencer & Co. had expanded and protected their business by the design (and patenting) of new improved spring designs every decade, Spencer Moulton maintained close links with the Chief Mechanical Engineers (CMEs) of these railway companies. Foremost among these was Sir Nigel Gresley, CME of the London and North Eastern Railway (LNER), designer of the 'A4 Pacific' locomotive class – the most famous example being 'Mallard' which in 1938 set the world speed record – 126 mph – for a steam train, a record that stands to this day. Behind these streamlined A4 locomotives trailed rakes of luxurious varnished teak carriages; high-speed comfort and safety being assured by the double-bolster compound 'Gresley Spencer Moulton' bogies. Conceived by Alexander Spencer as early as 1909, these bogies were refined by Gresley over a period of years and were fully developed by the 1930s.

A tale is told of how two eminent American rail engineers took the night sleeper from King's Cross to Edinburgh and were surprised and delighted to sleep soundly and not be woken at frequent intervals by the train riding roughly over points and crossings. On investigation, they found that their uninterrupted slumber was thanks to the unusual design of the carriage bogies; the eventual result being that several American and Canadian railway companies specified Gresley Spencer Moulton bogies on their trains. Astonishingly, some of these bogies were still in normal service on British railways in the 1990s, and indeed their smooth ride can still be experienced on heritage lines – notably the Severn Valley Railway.

Despite his position as Chairman (having succeeded John Moulton in 1919), and his advancing years, Alexander Spencer continued his innovative work into his 70s. On his retirement, Alexander had no less than 177 patents to his name. His departure represented the end of the direct involvement of the Spencer family in the design and development of new products – this mantle would fall to the railway engineers Richard Thomson Glascodine (a key figure in the development of the use of rubber in railway applications who had worked with Alexander Spencer for many years) and his son Richard Kenneth. The First World War had, as noted earlier, cruelly taken Alexander's only son and heir apparent, George.

THE SCENE IS CHANGED

No. 1.—INDUSTRY.

THE much-talked-of period of Industrial Revival is at hand, and although the Transition Roadway is still blocked by many difficulties, the dawn is already red with much promise. In this revival Motor Transport is destined to play a big part. The war has forcibly brought home to us the supreme value of mobility. In our industrial activities its importance is equally paramount. The efficiency of the Commercial Vehicle is made or marred by the quality of the Tyres.

WOOD-MILNE SOLID BAND TYRES

assure the best possible service at the lowest mileage cost. They are manufactured by a firm whose name in the Tyre World is a guarantee of reliability. Our services are at your disposal, let us help you solve your after-war transport problems.

Write us to-day for List No. S12.

WOOD-MILNE, LTD., and GEORGE SPENCER, MOULTON & Co., Ltd.

Head Offices: 42-46, Wigmore Street, LONDON, W.1.

Telegrams: "Wudmiln, Wesdo, London," *also* "Spenmoul, Wesdo, London." *Telephone: Mayfair 6610 (5 lines).*
Manchester: 21, Albion Street. Birmingham: 204, Corporation Street. Belfast: 13, Donegal Square. Bristol: 141, Victoria Street. Dublin: 69, Middle Abbey Street. Glasgow: 10, Waterloo Street. Leeds: 68, Albion Street. Newcastle-on-Tyne; Haymarket Lane. Liverpool: 31-33, Leece Street.

Spencer Moulton Wood Milne advert

Similar to how the grouping of railway companies increased efficiency and economies of scale, analogous efforts were made in other industries. In the case of rubber tyres, Spencer Moulton formed an alliance in 1919 with Wood-Milne, makers of rubber tyres and footwear (soles and heels) since 1896. The two companies were joined under a holding company – "The Federated Rubber Growers and Manufacturers Limited". Their in-house magazine, *The Winged Arrow*, was full of post-war ideology and hopes for a bright new future: "Amalgamation is the order of the day – just as the League of Nations is being formed to pool the forces of freedom and the moral arguments of mankind, and to narrow the circle of friction, so great firms, like Wood-Milne and Spencer Moulton, are founding 'Leagues of Enterprise' to pool the assets and methods of commerce, to lessen the area of friction, to curtail the waste of overlapping, to contract the centre of control, and to extend the circle of efficiency. Tennyson's dream of the 'Federation of Mankind' is taking shape in the federation of man's commercial interests, and, the trader is again the pioneer of progress, the missionary of the gospel of efficiency and material reform."

Despite the grandiose rhetoric and new products such as the 'Spenwood' tyre, neither company realized the benefits of their joint forces. Whilst Spencer Moulton remained profitable throughout the 1920s, Wood-Milne's position deteriorated rapidly and they were soon sustaining heavy losses. Another subsidiary, Federated Textiles – established to manufacture the cords and tyre carcasses for Wood-Milne, fared little better. By 1922, Federated Rubber Growers were not able to pay dividends to shareholders as trading losses mounted. Nevertheless, partially supported by the profits of Spencer Moulton, the group continued their descent into the abyss for another four years. In 1926 there was a delay in publishing the accounts as a re-organisation plan was schemed; this failed before it began and in 1927 the Directors were grateful to be able to accept a welcome offer for the footwear side of Wood-Milne. Although this sale did not include the factory at Littleborough, the Wood-Milne tyre moulds were transferred to the Kingston Mills in an attempt to reduce production costs. Eventually, and in hindsight inevitably, Wood-Milne's losses brought down Federated Rubber Growers & Manufacturers and the group was pushed into receivership in June 1928.

Spencer Moulton themselves, as a profitable subsidiary, were saved by a refinancing deal that saw the Westminster Bank loan them £300,000 (in 2015, adjusted for inflation, some £17 million) on debenture and provide generous terms for working capital. This allowed the company to continue trading profitably, although the repayments demanded of them were substantial even in times of prosperity – usually absorbing more than 50% of profits and of course taking precedence over any dividend payments to shareholders. Shareholders of Federated Rubber Growers were offered seven 5 shilling Spencer Moulton shares for every twenty £1 FRG shares they held; a little better than nothing, but they were perhaps lucky to receive anything at all. Those who worked at the Kingston Mills, almost the entire working population of Bradford, could continue as before and were very grateful to be able to do so. Remnants of the

TENNIS BALL REBOUND

Under the Regulations a ball when dropped 100 inches on a concrete slab must rebound not less than 50 inches or more than 58 inches. Every ball is dropped, and the rebound noted.

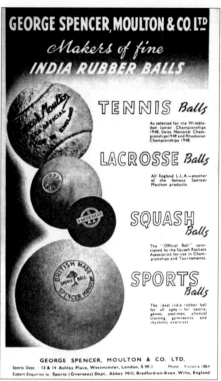

Spencer Moulton Tennis Balls. To ensure consistency, each ball was individually checked for weight and bounce height.

The Spencer Moulton sports goods department didn't only make tennis balls,as this advertisement showed.

A rare surviving box of Spencer Moulton tennis balls. (AMCT/DF)

Wood-Milne Tyre Co. were absorbed into the American tyre company BF Goodrich.

The tyre business was later to be largely transferred to the Trowbridge Tyre and Rubber Co. of Yerbury Mills; however Spencer Moulton maintained a storage depot in Trowbridge into the 1930s. John Moulton had resigned as Chairman in 1919 and had been replaced by Alexander Spencer; John died in 1925 and his elder son and sole heir (his younger son Eric having been killed during the First World War), John Coney Moulton, died just a year later. John Coney had served with the 4th Wiltshire Territorial Regiment in India during the First World War and in 1917 became a co-ordinator of Intelligence Services for the East, based in Singapore. Whilst having had no more than a cursory involvement with the rubber business (he was an entomologist), he did visit the Pundut Estates and is recorded as having met Sydney Spencer there in 1918. His passing left the Moulton family with no interest (other than shareholdings and ownership of the land) in the Kingston Mills - John Coney's three children were all of less than ten years of age.

As the tyre manufacturing plant was wound down, the Sports Goods department was born. Its principal product was the Spencer-Moulton tennis ball, which was adopted as the official ball of the British Lawn Tennis Association as early as 1932. The manufacture of such items was no easy task, for both weight and 'bounce' had to be measured and recorded within tight limits – every ball had to be tested. International tennis player Bill de Manby was employed at the company's London office and it was largely thanks to his endorsement – and his contacts – that the Spencer-Moulton tennis ball proved so popular at all levels of the game. There were also lacrosse balls, squash balls, hockey balls, golf balls, racquet grips and other sporting items.

Whilst durably popular, the sports market was very much a seasonal business, and indeed that depended on a season of good weather. Thus in 1936 the Chairman lamented the poor weather prior to Whitsuntide and the impossibility of making up for the loss of sales later in the year, cautioning that they must accept such a business would fluctuate according to the vagaries of the weather. Golf balls, however, were to be considered uneconomic to manufacture and were dropped from the product range. The Sports Goods department was based in the Abbey Mill in Church Street and survived for twenty years before being axed in 1953.

As demand for railway mechanicals continued to increase – and with the production site at the Kingston Mills now being rather full – subsidiary companies were formed for local manufacture in France (1929), India (1934) and Australia (1935). Despite this expansion, trading conditions in the 1930s 'Great Depression' were not favourable and losses were recorded in 1931 and 1932, putting the company into default on its debenture repayments. The next three years saw a return to profitability but scarcely enough to repay the agreed terms (£25,000 per annum) with the Westminster Bank; shareholders would have to wait, as they had been doing since 1930. By 1935 the company was experiencing an acute cashflow crisis. As the number of unpaid bills mounted, the difficulty in maintaining full production increased as suppliers became

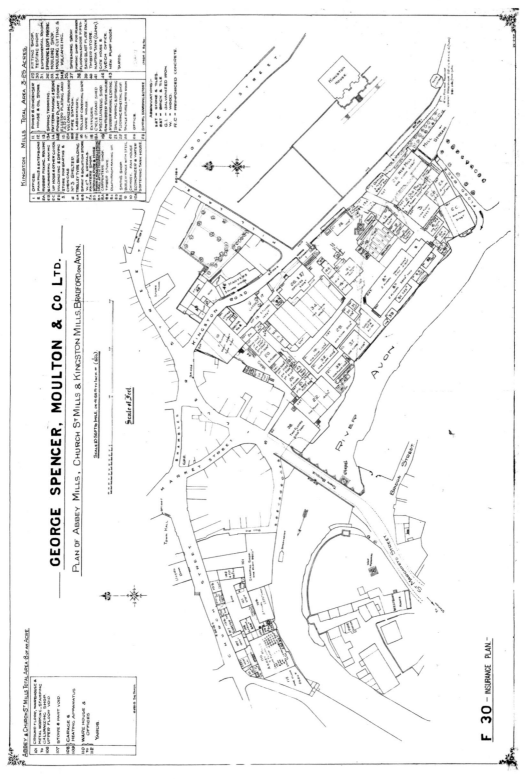

1940 Insurance Plan - note how the Spencer Moulton rubber works - Kingston Mills and Abbey Mills - completely dominate the town centre. (AMCT)

Here are some of the firms and Service Ministries for whom we carried out work during the war.

THE ADMIRALTY.
MINISTRIES OF
    SUPPLY
    AIRCRAFT PRODUCTION, AND
    WAR TRANSPORT.

AIRSPEED LTD.
ALBION MOTORS LTD.
ARMSTRONG WHITWORTH AIR-
    CRAFT LTD., SIR W. G.
AUSTIN MOTOR CO. LTD., THE
AVELING-BARFORD LTD.
AVIMO LTD.

BLACKBURN AIRCRAFT LTD.
BOULTON PAUL AIRCRAFT LTD.
BRISTOL AEROPLANE CO. LTD.,
    THE
BRITISH OVERSEAS AIRWAYS
    CORPORATION

COLE, E. K., LTD.
COSSOR, A. C., LTD.
COUNTY OF LONDON ELECTRIC
    SUPPLY CO. LTD., THE
CROSSLEY MOTORS LTD.

DAIMLER CO. LTD., THE
DE HAVILLAND AIRCRAFT CO.
    LTD., THE
DENNIS BROS. LTD.

ENGLISH ELECTRIC CO. LTD., THE

FAIREY AVIATION CO. LTD., THE
FERRANTI LTD.
FORD MOTOR CO. LTD.

GENERAL ELECTRIC CO. LTD.,
    THE
GRAMOPHONE CO. LTD., THE

HANDLEY PAGE LTD.
HAWKER AIRCRAFT LTD.
HUMBER LTD.

JACKMAN, J. W. & CO. LTD.

LONDON AIRCRAFT PRODUCTION
LISTER, E. A. & CO. LTD.

MARCONI'S WIRELESS TELEGRAPH
    CO. LTD.
McMICHAEL RADIO LTD.
METROPOLITAN CAMMELL
    CARRIAGE AND WAGON CO.
    LTD.
METROPOLITAN VICKERS
    ELECTRICAL CO. LTD.
MILES AIRCRAFT LTD.
MORRIS MOTORS LTD.
MORRIS COMMERCIAL CARS LTD.

NAPIER, D. & SONS, LTD.
NUFFIELD MECHANISATIONS LTD.

PERCIVAL AIRCRAFT LTD.
PETRO-FLEX TUBING CO. LTD.,
    THE
PLESSEY CO. LTD., THE
POWER JETS (RESEARCH AND
    DEVELOPMENT) LTD.

ROE, A. V. & CO. LTD.
ROLLS ROYCE LTD.
ROOTES SECURITIES LTD.
ROTAX LTD.
ROVER CO. LTD., THE

SAUNDERS-ROE LTD.
SCAMMELL LORRIES LTD.
SHORT BROS. (R. & B.)
SHORT & HARLAND LTD.
SIEMENS BROS. & CO. LTD.
STANDARD MOTOR CO. LTD., THE
STOTHERT & PITT LTD.
STURTEVANT ENGINEERING CO.
    LTD.

THORNEYCROFT, JOHN I. & CO.
    LTD.
TYRESOLES LTD.

VAUXHALL MOTORS LTD.
VICKERS-ARMSTRONGS LTD.
VISCO ENGINEERING CO. THE

WESTLAND AIRCRAFT LTD.
WOLSELEY MOTORS LTD.

24

From 1939 to 1945 Spencer Moulton's production was dedicated to the war effort. Here are some of the companies that relied on rubber parts from Bradford on Avon. (GSM) nervous. Eventually a moratorium was agreed which allowed for continuity of production, equal treatment of creditors and a committee of creditors having direct access to the Spencer Moulton Board of Directors. An agreement was reached with the Westminster Bank to re-structure the 1928 debenture over a longer period, to allow the payment of dividends, and for the Bank to extend overdraft facilities to pump more cash into the company.

The Second World War brought much greater changes to Bradford on Avon than the First had. Rather than trench warfare, battles were also fought in the air and on the sea – and in the laboratories and factories. Aside from the widespread use of railways for supplies and troop transports, rubber played an increasingly important role in aviation and land transport, from engine mounts to seals, undercarriage to tyres. There was also the manufacture of gas masks and inflatable boats. Spencer Moulton worked shifts – usually 24 hours a day – and extraneous lights had to be kept to an absolute minimum. Whilst the Kingston Mills did not appear to be on the Luftwaffe's list of targets, Bradford was uncomfortably close to the flight paths to Bristol which most definitely was. Bomb shelters were built and used, but gradually the pressure of work increased and air raid sirens went unheeded. Only a 'red alert' from Budbury would cause the workers to make for the shelters.

As well as evacuee children, others escaped the capital for the duration and found new places to live in Bradford. These included Jack Spencer and most of the staff of Spencer Moulton's London offices. Jack and his wife Angela set up home in an apartment in Manvers House; downstairs became the temporary headquarters of the company. Bradford itself was not detached from military activity, as Britain's secondary defence line was routed along the river Avon and the Kennet & Avon canal. In the event of a German invasion establishing a foothold on the south coast, British troops would have withdrawn to this final defence line. The Kingston (and Abbey) Mills sat right on this makeshift fortification. Material to create a blockade over the town bridge was put in place; fortunately this was never needed and the only threat came from the air.

As in the First World War, around a quarter of the workforce were called up – 183 men and 14 women. Most of the positions they vacated were filled by local women and production was rarely affected by lack of staff; getting hold of enough raw material was the primary concern. The Chairman's report in 1943 was a masterpiece of understatement:

> The difficulties inseparable from war time conditions with which we had to contend last year are still existing and are not likely yet to decrease;
> Our output will continue to be governed by the amount of raw material available and I fully expect that, due to this cause, there may be some falling off;
> As regards our rubber estates in Malaya, now in occupied territory, no information is available.

With greatly increased demand, and with no raw rubber available from the Pundut Estates and precious little elsewhere, 'Dr' Pickles introduced synthetic rubbers to the works, chiefly 'Government Rubber – Styrene' or 'GR-S' for short. GR-S, a co-polymer of Styrene and Butadiene, was developed in the USA in the 1930s and had many similarities with natural rubber – the major difference in finished product form being that whilst natural rubber softens with age, GR-S hardens. For the manufacturer, it brought new challenges in compounding and mixing as well as differences in mechanical properties, 'in-mould' times and moulding temperatures.

Spencer Moulton were producing over a million items a month for the war effort, including items as diverse as sea fuelling hose, hot oil joints for aircraft, fin dampers, cylinder seals, electrical parts for rockets, rubber mouldings for radio sets and transformers, sealing rings for tanks, suction hose, and seals for depth charges and mines. Owing to their proximity to the Bristol Aeroplane Company, the creative and experienced minds at Spencer Moulton were frequently asked to help with their specific requirements, but this was only a small part of their overall workload. A list (which in itself is not exhaustive) of wartime customers is reproduced here, demonstrating the reach of Spencer Moulton products into all areas of manufacture. All of this was in addition to essential items for railways, which were seeing service far in excess of peacetime usage.

7 Post-War Ambition

"Our past has been founded on the application of rubber to engineering,
especially railways. We are confident that the future holds even wider
scope in the same field of specialised technical application of rubber;
and to this end we are devoting our efforts"

Alex Moulton 1952

As the outcome of the war began to look more positive, Spencer Moulton was keen
to look to the future and to expand their manufacturing facilities even further.
Those who know Bradford will appreciate that the Kingston Mills site lies close to
the centre of the town, bounded by the river Avon on the south and east sides and
The Hall Estate to the north. To the west is the town centre itself. The only option
to expand the main site lay in the paddock, part of The Hall Estate but south of the
mill stream that fed the Kingston Mill. This land belonged to the Moulton family,
and following the death of Alice Moulton in 1941 this, and the freehold to the whole
Kingston Mills site, passed to the three children of John Coney Moulton – John, Dione
and Alex. Somewhat to the surprise of them all, Jack Spencer floated the idea of
the company buying the freehold to the whole site, including the paddock, and this
offer was accepted after a little adjustment to the suggested price. Whilst it appears
that Dione took her share of the sale price in cash – £10,000 – John and Alex were
already looking to become more involved at Spencer Moulton and took a significant
portion in shares. Jack Spencer may have suggested this himself, as the company
was not really in the position to pay the full sum, or indeed anything approaching
it, in cash. Alex Moulton joined the company in 1945 as an assistant to Jimmy
Chrystal, the Works Manager. Chrystal was highly intelligent and straight-talking; he
was respected and liked in equal measure. Before long, both Chrystal and Dr. Pickles
urged Alex to return to Cambridge to finish his degree in Mechanical Sciences - his
studies had been interrupted by the outbreak of war shortly after completing his first
year at Kings College.

Railways played a much greater part in the Second World War than in the First, and
the combined effects of national economies being on a 'total war' footing, the greater
utilization of track and rolling stock, and the six year duration left railway companies
with enormous backlogs of maintenance . 'Make do and Mend' could not be extended
much further in the case of buffer and bearing springs and the Spencer Moulton
factory was soon back to full capacity. The spring shop often worked 24 hours a day,

busily supplying 80% of the world's railways. The company returned to profitability and was able to pay reasonable dividends.

As railway modernization programmes were being established around the world, demand increased further as new rolling stock was built, and Spencer Moulton was well-placed to fulfil this. The standard 'concentric ring' buffer spring, a Spencer

A

⇥SPENCER MOULTON⇤

INVENTION

STANDARD WAGON
BUFFER SPRING.

DESIGNED and PRODUCED

by GEO. SPENCER MOULTON & CO., LTD.
2, CENTRAL BUILDINGS, WESTMINSTER, S.W.1.

The archetypal Spencer Moulton buffer spring, the '471'

Concentric buffer springs being removed from the heated vulcanizing chamber.

Moulton design, was in almost universal use and the company had in their possession vast amounts of tooling to manufacture this spring in many different sizes. The only problem that could arise was aggressive price competition from other manufacturers, as the patent protection on these designs had expired. Previously they had almost had the field to themselves, but the size of the supply contract offered by the newly nationalized British Rail was to tempt other players into the market. The standard buffer spring specified by the British Transport Commission was of the Spencer Moulton type, but given the patent position any rubber manufacturer could supply them. Fortunately Spencer Moulton's reputation for quality prevailed in these early post-war years.

In the late 1940s, the 'Technical' base of the company began to shift from the London Office (Ashley Place) towards Bradford on Avon. It will be remembered that George Spencer & Co., which had largely supplied the designs which Moulton manufactured, were London-based; and this somewhat inefficient split had remained in place. The design and technical work took place in London, the manufacture in Bradford on Avon. A major factor in the move westwards was the retirement of the last of the Glascodines – Richard Thomson Glascodine had been appointed Spencer Moulton's Technical Manager in 1920 (replacing Alexander Spencer who had risen in position to Chairman) and Technical Controller (at the age of 70) in 1939. His son, Richard Kenneth Glascodine, worked with him for many years and retired shortly after his father's death. The Glascodines were traditional railway engineers and had continued the pioneering work of the Spencer family in the design and development of rubber compression springs that were more effective, cheaper to produce and more suited to the high speeds and heavy train loads of the era. They also handled all of the intellectual property, primarily patents, for the company. No qualified Technical Manager was ever appointed to succeed them; and without this, the significance of the London office was greatly reduced.

The Head Office, however, remained at Ashley Place. Seemingly rooted in another age, Jack Spencer (Chairman) and Geoffrey Godfrey (Managing Director) sat opposite each other at a Victorian partners desk. The Company Secretary and Chief Accountant, Henry Lambert, had his own office. Frank Evelyn-Jones, the solicitor and architect of the deal that rescued the company from the Wood-Milne fiasco in the 1920s, appeared only at board meetings. One can only imagine the difficulties in managing such a large company from a distance. It was seen as essential to maintain a presence in the capital, but with the Technical Department being reduced almost to an administrative function it seems strange that the power base remained there for so long.

In a parallel development, a new building was being erected in Bradford on one of the very few remaining open spaces on the site, directly opposite Manvers House. This, the 'Centenary' building, was completed in 1948; one hundred years after Stephen Moulton opened his rubber factory in the Kingston Mill. The new building was to be

The Opening of the Centenary Building in 1948. (AMCT/DF)

Spencer Moulton Centenary Garden Party 1948: Chairman Jack
Spencer addresses the assembled gathering. From left to right:
Geoffrey Godfrey, Beryl Wood (Alex Moulton's Mother),
Dr Pickles, Jack Spencer, Angela Godfrey, Alex Moulton,
Trade Union representative, Pheasant Lambert. (AMCT/DF)

devoted to research into new rubber products, new processes, and new compounds – the primary intent being to seek out possible applications of the use of rubber as a springing medium. Of course, rubber springs had been the mainstay of the Kingston Mill for nearly a century, but these had been 'simple' compression springs. Indeed, the working knowledge of rubber as an engineering element was still limited. Whilst its use and behaviour in compression was widely understood and accepted, many engineers distrusted the material enough to not specify it where static load bearing was required (for example as a primary bearing spring on a railway carriage), for fear of it taking a permanent 'set'. The management were mindful of the fact that the way to stay ahead of the competition was to innovate, just as George Spencer had done so ably in the nineteenth century. The aim was develop new products that would put Spencer Moulton back at the forefront of technical progress in the field of rubber engineering.

The celebration of the centenary was both enthusiastic and dignified. As well as the opening of the Centenary Building, there was a grand garden party at The Hall. A booklet – *A Hundred Years of Rubber Manufacture* - was distributed to staff and other interested parties. This, largely drawing on family archives, covered the origin and the setting up of the works and a little detail on current practice. At the end of the booklet it lamented that the lack of space did not allow the mention of the long line of managers, foremen, forewomen and work-people that had spent much of their lives in the service of the company.

Spencer Moulton Centenary Garden Party 1948: The Spencer Moulton Brass Band
(in existence for generations) entertain the crowd in the grounds of The Hall. (AMCT/DF)

This may be a case of 'better late than never', but it is worth noting that Spencer Moulton was a good and respected employer, and the aim of most school leavers was to work at the Kingston Mills. Several families worked there for generations – the Venells, Holbrooks, Panzettas, James', Damsells, Nokes', Mayalls, Bigwoods and many more. One Ronald Watts joined the company in 1933, his father Henry had joined in 1905. His grandfather, Eli, clocked no less than 52 years; and great grandfather Watts was one of Stephen Moulton's first employees. The Venell family were even more remarkable – Thomas, John and Mary Venell all started work for Stephen Moulton in 1850, and their direct descendents recorded a combined service of nearly 900 years. Bob Venell first clocked on in 1870 (at the age of ten) and finally retired in 1932. Trade Unionist Leonard Viles' comments on Spencer Moulton echoed the feelings of many: "I have been here 37 years. I am interested in my work and have no complaints. There is none to whack this firm." Nelson Banks and Cecil Bray, with 90 years service between them, added in unison "this is a happy family concern, and second to none in the world". A reporter from the 'Wiltshire News' concluded that "it is a tonic to see such an immeasurable degree of free co-operation between the worker and employer." Perhaps the last word on this should go to Walter 'Woppit' Holbrook, who retired in 1960 after 54 years at the Kingston Mills: "Some nights I dream that I'm working back there, then I wake up – disappointed."

Inside the new building, the work of the research department was largely concerned with using rubber in different ways – following the development of reliable rubber-to-metal bonding, a whole new field of properties and possibilities for the uses of rubber in engineering could be opened up. For the first time, one could design rubber springs that could be loaded in torsion, or in shear (or a combination of the two). Compared to compression springs, and given suitable designs, there was great potential in economy of material, weight and cost. Formed in 1947, the new research team – a small group with diverse skills – was led by Alex Moulton, great-grandson of Stephen and youngest son of John Coney Moulton. Their greatest challenges were to make mathematical sense of rubber when loaded in these new ways – hitherto this had just been guess-work – and to establish rules to ensure adequate fatigue life of torsion and shear springs. There was also the question of quantifying and reducing 'creep'; the permanent deformation of the rubber under static loads. The man largely responsible for the team achieving these aims was Philip Wilson Turner, a brilliant physicist and mathematician whom Alex had recruited from the University of Cambridge – Turner had been one of his Mechanical Sciences tutors.

Aside from the increase in demand from railway companies' post-war repair, renewal and modernization programmes, the automotive industry was embarking on an enormous expansion that would transform the motor car from a play-thing of the aristocracy into a modern, mass-market mode of transport. Their position could be compared to that of the railways almost a century before – on the cusp of explosive growth. George Spencer & Co., and the Moulton Rubber Co., had built their businesses on the growth of the railway industry; now Alex Moulton and the directors at Spencer Moulton were determined to re-establish their eminence in the rubber world by developing rubber technologies for use in automobiles. Even the technological position was analogous: vulcanized rubber was invented after the coming of the railways but before the expansion phase; and the rubber-to-metal bonding process was developed shortly before the rapid growth in car manufacture. This latter technology was deemed essential to allow the use of rubber in automotive suspension, as if rubber is used in compression the 'install envelope' – space needed for the spring – is very large for a given static deflection. A road vehicle would need a prohibitively tall compression spring to achieve a sufficient amount of wheel movement to cope with typical roads. By contrast, a railway carriage requires a relatively small deflection spring by virtue of running on smooth rails, and rubber in compression is entirely suitable.

A torsional shear rubber spring, the 'Flexitor®', was patented by Spencer Moulton in 1948. By 1949 a range of different sizes of Flexitor (up to 2.5 tons axle load) were being tested at MIRA for use on trailers and caravans. Marketing was undertaken with the co-operation of trailer parts supplier The Bramber Engineering Co. and the Bradford-manufactured units were soon to be sold in substantial quantities in the UK and overseas. Of primary interest to trailer and caravan manufacturers was the simplicity of fitment, reduction in maintenance requirements and the very small install envelope enabling a lower platform height to be achieved. The success in trials and the great deal of interest generated led to the formation of a subsidiary company, Spencer Moulton Flexitor Ltd. The Chairman was Jack Spencer, with Alex Moulton as Managing Director and John Moulton (eldest son of John Coney Moulton and Alex's older brother) as Company Secretary. The Technical Department in London was finally closed, with two of the remaining draughtsmen being transferred to Bradford and the upper floors of the London office being let to another company.

It is apparent now that a driving factor in the creation of the Flexitor Co. was to free the Research Department, and in particular Alex Moulton, from the strict rigidity of rules and procedures that beset large organisations. Without doubt the Research Department had been very active in pursuing commercial leads without the involvement of Spencer Moulton sales and marketing staff, and this inevitably led to friction in some areas. Geoffrey Godfrey (Joint Managing Director with Jack Spencer) certainly found Alex difficult to handle. He was not the first, or indeed the last, person to find this. The situation was made more delicate by Alex's position as a substantial shareholder, and no doubt by his name and relation to the founder. Shortly after the founding of the Flexitor Company, Godfrey abruptly resigned from the firm; he was replaced by Alex who became Assistant Managing Director. Very soon afterwards, in a move that raised questions of impropriety in the industry and in the press, Godfrey took up a position as Managing Director in the brave new modernist world of the Brynmawr rubber factory.

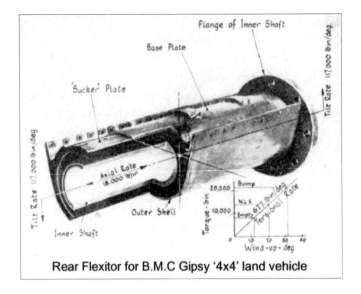

Rear Flexitor for B.M.C Gipsy '4x4' land vehicle

Spencer Moulton Flexitor
Suspension Unit for Austin
Gipsy 4x4 vehicle. (AMCT)

Completed Spencer Moulton Flexitor
rubber-in-torsion springs are stacked
for packing and despatch. (AMCT)

With the objective of a large-scale manufacturer adopting the Flexitor for use on passenger cars, several projects were initiated to demonstrate the advantages of the Flexitor suspension over the conventional. These included 70,000 miles of running in a Bristol 404 luxury car and, at the other end of the scale, the Flexitor was adopted as rear axle springs for the Bond Minicar and several thousand were produced. It was not until 1955 that the Austin Motor Company asked Spencer Moulton and the Flexitor Co. to propose a suspension system for a cross-country vehicle designed to compete with the Land-Rover. This project was successful and volume production of suspension units for the Austin Gipsy commenced in Bradford on Avon in 1958, although in rather different circumstances than originally envisaged.

It is interesting to note that, in principle, the torsional shear spring devised by Spencer Moulton was not new. In America, Alvin Krotz had developed the 'Torsilastic' spring and this was the standard fitment on Greyhound buses. Alex Moulton was aware of Krotz's work and the Flexitor was devised to both improve on Torsilastic and to avoid infringing on existing patents. Moulton travelled to America to meet Krotz, and any potential query on infringement was avoided by the remarkably simple (and common sense) provision of a non-exclusive 'cross' licence. Thus either party was free to use the other's patent in their territory. There can also be little doubt that the descriptive name 'Torsilastic' influenced Moulton in the naming of his inventions. This creation of new words -"a new thing needs a new name ... if something has not existed before it cannot be familiar and therefore any description must either be vague or lengthy" – was welcomed by his trade mark attorneys, although they were a little frustrated by their illustrative nature..

Amongst the miscellaneous outputs from the research department came a few developments of great significance. One was facilitated by chance of geography – the Naval Constructors and the Engineering Branch of the Admiralty were both based in Bath, with many of the officers living in Bradford. Hence the Flexitor Co. were contracted to design an anti-vibration mount for diesel generators installed in wooden-hulled high speed craft. These generators were huge – the design load per mount was 600lb and a later version was rated at 1000lb. These were produced in large quantities and the bonded rubber cone design was later copied by many engine manufacturers. Another involved the conversion of a contemporary Morris Minor to a rubber suspension system. Whilst the technology used was at a rudimentary state of development, it demonstrated what could be achieved well enough to excite the mind of the rising star of British automotive design, Alec Issigonis.

Issigonis, who had established his reputation par excellence with the design of the Minor, was initially dismissive of rubber as an engineering material. After the rubber-suspended Minor covered 1,000 miles on the Motor Industry Research Association's pavé (cobblestone) track without any failures (a coil spring suspension could only manage half that), he was convinced that rubber suspension was the way forward; thus Alex Moulton and the Flexitor Co., became involved in the design of the suspension for

Issigonis' latest project, the Alvis TA350. Issigonis made frequent 'stimulating and lively' visits to the works at Bradford and the collaboration between him and Alex Moulton was to have a profound effect on car design for a generation.

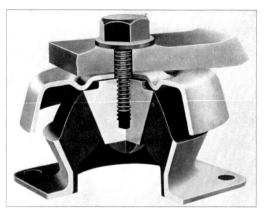

The 'Admiralty Mounting' compression and shear rubber engine mount. Later known simply as the 'cone spring' and fitted (in a larger version) on millions of BMC/Leyland/Austin-Rover Minis. (AMCT/DF)

In a gross over-simplification of a story that is perhaps better told elsewhere, a version of the 'Admiralty Mounting' – a rubber cone engine mount that cleverly combined stressing the rubber in both shear and in compression – was re-purposed as a strut suspension spring (two cones in series 'nose-to-nose' integrated with a telescopic damper) on the Alvis prototype. From this came the discovery that a rubber cone could be used as a fluid displacer as well as a primary spring, and the cones on the front and rear could be hydraulically interconnected to reduce the 'pitch' motion of the car. Alex would later recall "I shall never forget the revelation at Coventry, with Alec (Issigonis) driving, of experiencing the 'big car' ride due to the lowered pitch frequency. The reality of the benefit of fluid interconnection was thus revealed and the seed was sown, not that we realised it then, for a radical new suspension to be made in vast numbers." This was the first physical embodiment of what would become known as Moulton 'Hydrolastic'® suspension, later to be fitted to 4 million cars from 1962 to 1978 – Austin/Morris 1100,

Alex Moulton and Alec Issigonis discuss the Hydrolastic suspension unit, the interconnected hydraulic rubber suspension system for the new Morris 1100. Initial development work was undertaken by Spencer Moulton and Alvis cars. (AMCT)

1300, 1800, Mini and Maxi; and Moulton 'Hydragas'® continued until 2002 when the last MGF was produced. The intention was for Hydrolastic to debut on the new Morris Mini Minor / Austin Seven in 1959 but the development programme was not sufficiently advanced to allow this and the cone spring was used instead. Certainly the tiny install envelope was an essential component in the marvellous space utilization of the Mini. Hydrolastic made it onto the Mini in 1964 only for the cone spring to return in 1970; and between 1959 and 2002 Moulton suspension found its way under all four corners of nearly thirteen million British cars.

Other than prototype versions, Moulton Hydrolastic, Hydragas and cone springs were not manufactured in Bradford on Avon, but all the development and testing was completed by Spencer Moulton (and its Flexitor subsidiary) and, later, by Moulton Developments at The Hall. Most of the production was undertaken by Dunlop in Coventry and, at their peak in the 1960s, 1,200 staff were employed in the manufacture of Moulton suspensions – more than Spencer Moulton had employed at any time in their history.

Returning to the 1950s, we can see that whilst the Research Department and the Flexitor Co. were building a new future, the parent company was enjoying – somewhat fleetingly – the post-war revival in demand for railway mechanicals. In 1952 they posted a healthy profit of £107,000 and paid generous dividends of 10%. This prosperity was not to last. With trading conditions becoming increasingly competitive, the corresponding figures for 1953 were £25,000 and 5%. An extensive reorganisation of the whole of the company's operations was planned, to include a widespread modernization of the premises, plant and machinery as well as manufacturing and administrative processes. It had become clear that, despite the overall profitability, the company had little grasp of actual costs even on a departmental level. Some obvious economies were made, such as the cessation of 'sports goods' production and the closure of the Technical offices in London, but it was not until the departure of Godfrey and the retirement of Lambert in 1953 that real efforts were made to introduce modern methods of accounting and costing that would allow proper measurement of performance and profitability. This was not before time, as 1954 saw a further slide in the company's fortunes.

Spencer Moulton buffer springs being despatched from the Abbey Mill. In times of high demand, the factories would run 24 hours a day. (AMCT)

8 Hitting the Buffers

"Your reckless policies have ruined our poor little company. Please leave
now."

Alex Moulton to David Granville, 1955

In advance of his resignation, and in order to bolster the financial mettle available to
the company, veteran Director Evelyn-Jones proposed the appointment of Charles
Woolveridge to the Board; he joined in early 1954. Woolveridge, a Director of the
accountancy firm Binder Hamlyn, was alarmed by the lack of reliable information
available to the Directors and suggested that a Management Consultant (from his own
company, naturally) should be engaged to look into, and report on, the manufacturing
operation in Bradford. Evelyn-Jones left as David Granville, the consultant, arrived
at the Kingston Mills. At around the same time, John Moulton joined the Board as
Company Secretary.

Granville was full of energy and determined to make his mark; he soon set out plans
for major re-organisation, including wholesale changes to long-established methods of
working. Massive cost savings were promised, along with streamlined production
and a smaller workforce. The numbers looked appealing to the Board, and Granville
– perhaps thinking that he had gained their confidence and respect – intimated that,
if offered the position of General Manager, he would resign from Binder Hamlyn.
He was gently reminded that it was correct and proper to stick to the job that he
was brought in to do, and to leave the executive functions to the existing Board and
management.

The most critical question facing the company at the time was that of pricing,
specifically what price should be quoted for the all-important British Rail spring contract
for 1955. The loss of this work would be disastrous for Spencer Moulton, as it
formed the bulk of their business. As mentioned earlier, the value of this contract was
such that it had attracted other companies to quote for supply – amongst them were
Dunlop, Clyde and Avon Rubber. Price competition was fierce, but Spencer Moulton
needed this contract desperately. Granville introduced the Board to the latest trend
from American Business Schools – 'marginal costing'. It was decided that the price
tendered could be based on marginal costing, with the expectation that economies of
scale would provide the profitability.

Granville calculated the price for the standard '471' spring, and the consensus was
that no other company would be able to match it – and so it proved. Spencer Moulton,

much to their relief, won the bulk of the contract and the future of the spring shop – the core of the factory for a hundred years – seemed assured. But – as was soon to be discovered – there was a major problem. Either the concept of 'marginal costing' was not correctly understood or applied, or the costing data itself was inaccurate in the first place. The price quoted for each spring only just exceeded the direct cost and made no contribution to overhead recovery (heat, light, insurances, buildings etc.) at all. In effect, each spring was being sold at a loss, so the more that were sold, the greater the deficit. On discovering this, Jack Spencer attempted to re-negotiate the contract with British Railways but with no real result.

Alex Moulton, characteristically, was more direct. At the next meeting in Manvers House he cornered Granville and, with no preamble, addressed him quietly with cold fury in his voice – "Your reckless policies have ruined our poor little company. Please leave now." Granville cleared his desk and left; he did not return to Bradford again. Alex reported his action to the Board and it was met with universal approval. In hindsight this was perhaps a little harsh on Granville, certainly the Directors were guilty of a lack of financial rigour in accepting his calculations without seriously questioning the data or the method.

The financial position at Spencer Moulton quickly descended from merely loss-making to disastrous; a long-term solution to the problems that beset the company had to be found and, this time, the Westminster Bank were not so obliging as before. Alex Moulton – being aware that he had rather more than most to lose in this debacle – invited Charles Floyd, Chairman of the Avon India Rubber Co., to lunch at The Hall. Floyd was married to Mary Fuller, granddaughter of George Fuller MP who had been Avon India Rubber Chairman at the end of the 19th century; Floyd represented the continuing interest of the Fullers in Avon.

Avon, and Floyd, were keen to tap into the worldwide reputation that Spencer Moulton enjoyed in the railway industry and to exploit some of the advanced rubber engineering research and depth of skills and experience that was available in Bradford. Spencer Moulton needed new management and new markets – and new financing. Clearly there was enough synergy for a merger or takeover to be mutually beneficial, and so further discussions were forthcoming. These took place in utmost secrecy so as not to start rumours amongst the workforce or on the Stock Exchange as to the future of the company. It was not until the factory summer shutdown in August that a small team from Avon were able to look around the factory in Bradford.

Private meetings between Directors and managers of both companies continued into the autumn of 1955. By this time, Spencer Moulton were dealing with an increasingly impatient Companies House with reference to their 1954 accounts which were seriously overdue for publication. Mindful that any qualification on these accounts would adversely affect – or possibly ruin – any deal with Avon, Jack Spencer in particular worked hard to put the best gloss he could onto the situation. Eventually they were approved, leaving the Spencer Moulton Board with not much to do other than to wait for Avon's decision, keep the factories open and spend as little precious cash as possible.

John Moulton was in attendance at a Board meeting at the London offices at the end of October when he was called to the telephone. On being advised that Mr Lovell, the Avon Company Secretary, was on the line he took the call 'with much apprehension'. Lovell's message was short and direct. Avon were prepared to make a non-negotiable cash offer of 6/6d per share for the outright purchase of George, Spencer Moulton & Co; he would send written confirmation forthwith and he expected them to accept the offer on receipt. John Moulton was able to relay this information, with a great sense of relief, to the Board in the next room. The offer far exceeded any of their expectations and the Directors confirmed acceptance – subject to shareholder approval – immediately.

The wheels were set in motion to complete the legal transaction on January 1st 1956, with a one-year transition period for some parts of the business, including the Flexitor Co. That this was achieved smoothly, with no notable problems, is testament to the outstanding co-operation of the staff at all levels in both companies. George Spencer, Moulton & Co held their 65th Annual General Meeting at The Swan Hotel in Bradford on Avon; thereby ending 108 years of independence as they began a new life as a subsidiary company of The Avon India Rubber Company.

Aerial view of the Kingston Mills Rubber Works. Note the swimming baths – gifted to the town by John Moulton senior – on the site of the present library (right of the town bridge).

(AMCT)

9 Avon in Bradford

The first public announcement of the putative takeover was posted in the national press on Saturday 10th December 1955: "The Avon India Rubber Co. Ltd. have offered to purchase the ordinary share capital of Geo. Spencer, Moulton & Co Ltd., general rubber manufacturers of Bradford on Avon, Wilts., and the Directors of Geo. Spencer, Moulton & Co. Ltd., are strongly recommending acceptance." The notice concluded with some notes on how the acquisition would shape the future for Avon: "It is Avon's intention to keep the company of Spencer, Moulton Ltd., in being, to continue Spencer, Moulton's policies of production and development and eventually to concentrate Avon and Spencer, Moulton general rubber production at Bradford on Avon. As an eventual result of this plan, it is hoped to concentrate tyre production at Melksham and to raise output by 50% by 1958."

So, far from being subservient to the Melksham headquarters, Spencer Moulton at Bradford was tabled to be the Avon group's factory for rubber mechanicals and other parts. Over 90% of Spencer Moulton shareholders agreed to accept Avon's offer and the business was duly concluded with Avon purchasing George Spencer, Moulton & Co. outright for a cash sum of £352,492. As well as the Bradford on Avon factory – Kingston Mills and Abbey Mill – Avon acquired the overseas operations of Spencer Moulton (albeit only 50% of Spencer Moulton France) including the Pundut Estates in Malaysia.

Whilst continuing as a separate company, there were changes at the top of Spencer Moulton: both Ashton-Cross and Woolveridge resigned. Alex Moulton had made it clear to Floyd that he would be leaving to start his own rubber engineering company but would stay for a year to manage the ongoing research and development at the Flexitor Co; and John Moulton stayed as Company Secretary for Spencer Moulton in 1957, taking an administrative position at Avon in Melksham after that. The new Spencer Moulton Board was jointly led by Jack Spencer and Avon director Oswald Swanborough, and day-to-day management would be the responsibility of existing Department Managers reporting to directors at Melksham. There was also no longer any need for the London office and the facility at Ashley Place was closed and sold in 1957. Jack Spencer continued to keep an office in London, even after relinquishing his executive duties in 1958. He retained his seat on the Board and indeed was elected President of the Avon group in recognition of his unparalleled experience, particularly in the 'railway rubber' field.

Floyd must have had mixed feelings when, at the end of the transition period, Alex

Moulton not only left but also took most of the Flexitor Co. and research department staff with him, including Philip Turner. Alex's new company, Moulton Developments – created to put rubber cone and Hydrolastic suspension into road vehicles (as described in chapter 7) – was formed as a triumvirate between Moulton as designer, Dunlop as manufacturer and the British Motor Corporation (BMC) as 'customer'. The inclusion of Dunlop rather than Avon was perhaps controversial; whilst Alex always insisted that Avon were "too small", I suspect that the real reason was that Dunlop, as well as being established as a major BMC supplier, were prepared to back him enthusiastically and unequivocally. Dunlop's Managing Director at the time, Joe Wright, is quoted as saying "We'll do absolutely anything, this is a wonderful opportunity. BMC are going to use rubber suspension – Goodness Me, you can have what you want." The split of the designs and intellectual property developed by the research department at Spencer Moulton remains unclear; certainly Avon retained the Flexitor rights and also marketed the cone spring, yet Moulton Developments also used the cone – most famously on the original 'Issigonis' Mini car. Avon continued to pay Alex Moulton an ex-gratia 'royalty' on Flexitor sales well into the 1960s.

Alex Moulton with a prototype rubber-suspended Moulton bicycle, and a Morris 1100 with Moulton Hydrolastic hydraulically interconnected rubber suspension. The 1100 would be the best-selling car in the UK in the 1960s. (AMCT)

Moulton Developments and Alex Moulton were also responsible for the design of the innovative Moulton bicycle launched in 1962. Featuring small wheels and suspension front and rear, this represented a total advance over the conventional bicycle in road-holding and ride. It revitalised the despondent cycle market to such an extent that by 1970, one-third of bicycles sold in the UK were Moulton-inspired. The front suspension used a composite rubber and coil spring arrangement; the rear unit utilised rubber loaded in compression and shear. The Moulton bicycle was produced in Bradford on Avon from 1962 until 1967; an improved version, dubbed 'The Advanced Engineering Bicycle', was launched in 1983 and is still manufactured in the grounds of The Hall. It is highly sought-after worldwide and is a respected design icon – in 2012 the architect Lord (Norman) Foster declared that "the Moulton bicycle is the greatest work of 20th century British design".

Aside from the Board and in the research areas, little changed at the Kingston Mills. Even the name – George Spencer, Moulton & Co. – was left unaltered, as their reputation with the railway companies was of great value. Gradually working practices were brought more into line with those at Avon's other sites and a degree of rationalization took place as common functions – research, technical development, administration – became group-based rather than site-based. There were to be no regrets and Avon remained very pleased with their purchase; Spencer Moulton reported a return to profitability in 1956 and thus were "fully justifying the confidence of the Board." Avon's total profit in 1957 was £371,448 – more than they paid for Spencer Moulton, even in a year where results were adversely affected by the Suez Crisis.

Whilst hitherto the Kingston Mills relied upon the traditional railway mechanicals business, the late 1950s saw many projects come to fruition. Some of these – the Flexitor suspension units for the Austin Gipsy, and the cone spring engine mountings – were initiated by Spencer Moulton. The Gipsy was launched in 1958, amidst a blaze of publicity extolling the virtues of its rubber suspension: "a greatly improved ride (especially in rough conditions) with maximum road-holding due to the four wheel independent suspension and special properties of rubber, which give instantaneous reaction to shock." Amongst other commentators, the Sporting Life was convinced – "It is the best springing system I have yet experienced. Suspension is really amazing. You can drive fast over rough country – and by that I mean off the track – in comfort."

Other newcomers, notably the inflatable boats, were Avon creations. Avon had manufactured inflatable boats in wartime but had since identified a market for them as yacht tenders. Four models formed the initial range – Redstart, Redshank, Red Crest and Red Seal. The Avon Inflatables department was based at the Abbey Mill and was an instant success – so much so that 'Avon' became almost a generic name for inflatable dinghies. Before long it became apparent that the Abbey Mill production was not sufficient to satisfy demand and production was moved to Dafen, near Llanelli in Wales, in 1964.

The Austin Gipsy with all-independent Flexitor rubber suspension. By the time the Gipsy was in production, Spencer Moulton had been bought by Avon Rubber. (AMCT)

Avon inflatable boats on show in 1962. The Avon Inflatables department was based at the Abbey Mill until 1964, when it moved to a larger facility in Wales.

The Abbey Mill was then re-purposed for car tyre manufacture – or rather re-manufacturing tyres, as these were remoulds. Apocryphal tales are told of how tyres were smuggled out of the mill for use on employees' cars. One trick was to fit inner tubes into the tyres, inflate them and cast them out of the window into the Avon. Timing was important, the aim being to get the tyres into the river half an hour before clocking off. A quick dash to Avoncliff after work and the tyres could be picked off the weir crest; an added bonus being that the operation could be concluded by a celebratory pint or two at the Cross Guns. Obviously, one would only go to such lengths on a day when the right size tyre for your car was being moulded. The less impecunious (and more honest) could take advantage of a generous discount (reportedly 35%) offered by the firm on new Avon tyres. The tyre remould business suffered from a surfeit of success and, in a similar fashion to the inflatable boats, was relocated to Bridgend in South Wales in 1968. For a time the truck tyre remould business stayed in the Abbey Mill before that too was ousted westwards.

Whilst Avon Rubber divested themselves of the Pundut Estates in 1961, Spencer Moulton continued to expand their facilities in Bradford. A new 28,000 square foot extension to the 'Paddock' building was opened in 1960 to further service the automotive and railway trade requirements. During the following year a new silicone shop was erected inside an existing building. Equipment included mixers, extruding and moulding machines, and curing; so that the manufacture of silicone items could take place under one roof. This was the first moulding shop to be equipped with heat extraction fans and air-conditioning units to prevent foreign matter from contaminating the product.

Overshadowed by the Abbey Mill, the Church Street Mill provided valuable production space for Spencer Moulton and Avon Rubber up until the 1980s. (WSA)

Following the departure of the remould tyre business, the upper floors of the Abbey Mill were converted into open-plan offices and a restaurant for staff. The ground floor was still retained for rubber manufacture – small aerosol gaskets and cup seals. These gaskets were die-cut from thin-rolled rubber sheeting, and the cup seals cut from extruded rubber tubing The business grew with the aerosol market in general, with Bradford supplying more than 50% of the world demand. A further extension to the Paddock was added in 1965 to help meet demand for these aerosol components and a rapidly-expanding line in moulded hoses for motor cars.

A new rubber store was also erected, and a new office block was constructed by the side of Manvers House to allow sales staff – those who sold products made at the Bradford site – to be moved in from Melksham. The new 'spacious and airy' offices were 'a great contrast to the difficult conditions which the office staff had to contend with previously' and the building was 'designed to fit in as far as possible with the existing factory buildings'. These offices still stand on Kingston Road; you can be the judge of how successfully the architects fulfilled this intent.

View westwards along Kingston Road in the 1970s.
The Centenary Building is in the centre, The Hall Estate behind the wall on the right.
(AMCT)

SRN-4 Hovercraft landing pads - the largest bonded rubber mouldings made
in the UK at the time. Note the oversize vulcanizing chamber. (WSA)

SRN-4 Hovercraft with Avon landing pads and skirt retainers.

Oswald Swanborough retired as Managing Director in 1966. His career at Avon had started in Bradford on Avon; for a few months in 1925 he had worked at the Greenland Mills before being transferred to Melksham. He had overseen Avon Rubber's ('India' had been dropped in 1963) growth from £2.2 million turnover to £26.6 million and he was to remain on the Board until 1970. He was replaced as Managing Director by his son, John Swanborough. Managerial talent clearly ran in the family as John represented the third generation of Swanboroughs who had been Managing Director at Avon. Avon Rubber saw many years of impressive growth, were active in the acquisitions market and were to remain resolutely profitable throughout the 1960s and into the 1970s. The stated aim of the Bradford on Avon factory being the centre for general rubber manufacturing remained, but the relentless growth of the company was to place increasing pressure on all the Avon group facilities.

Spencer Moulton – now part of Avon's 'Industrial Products Division' – continued to rise to rubber engineering challenges in diverse fields of application. In 1967 they met the biggest – and heaviest – request in their history. Aerospace and Marine Engineers Saunders-Roe were developing their mammoth 160 ton, 100 mile per hour, SRN-4 'Mountbatten class' hovercraft and needed rubber 'landing pads' to join the landing skids to the craft itself. With each pad weighing over 100 kilogrammes (and containing no less than 88 kilogrammes of rubber), these were the largest rubber-to-metal bondings that the Kingston Mills had ever produced and were believed to be the largest ever manufactured in Britain. Given the size of each pad, and the loading requirements – 80 tons in compression and 40 tons in shear – achieving complete and consistent bonding and vulcanization was both difficult and essential. As well as the units themselves, special test rigs were constructed to ensure that the specified load and deflection characteristics were met.

On a different scale but no less impressive was the contract that the Bradford on Avon facility won in 1968. Following four years of negotiations and extensive testing, Volkswagen placed an order for 40,000 quarter-light window seals for their famous Type 1 'Beetle' car. Having developed a new injection-moulding process to manufacture these seals, Spencer Moulton / Avon were able to reduce costs and improve quality in comparison to their competitors. Thus the Avon Rubber Group became the first British company approved as a supplier by Volkswagen: this contract remained in place for several years and many others followed in its wake.

By the time Chairman Charles Floyd retired in 1968, the relentless growth of the Avon empire was in full flight. Turnover trebled from £12.9 million in 1961 to £35 million in '68, and there was more to come. Seeking to expand their tyre business, Avon began construction of a new radial tyre factory in Washington, County Durham. In a foretaste of what the 1970s would bring, economic factors forced a change of plan and the new factory was sold to Dunlop before a single tyre had been made there. Avon re-trenched to west Wiltshire and re-affirmed their commitment to Melksham (tyres) and Bradford on Avon (mechanical and industrial products). Turnover continued to rise and eventually reached £53 million (with £2.2 million profit) in 1973. Avon was, by then, a very different company to the one that had taken over Spencer Moulton in 1956.

Catherine Hulbert sorting aerosol gaskets in the Abbey Mill. Peak production was *circa* 80 million gaskets per week, and they could often be found on the pavements in Church Street.

(WSA)

Production at Bradford reached new heights in 1969. Demonstrative of the worldwide reputation and supremacy of Spencer Moulton in the railway mechanicals business, significant orders were received from Finland, Belgium, Burma, Japan, USA, Sudan, Peru, East Africa, Malaya, Angola, Guyana, Iraq and Australia. The Technical team were taxed with another unusual problem on the SRN-4 hovercraft and were able to provide a highly creative solution. Despite being fabricated from tungsten steel, the skirt retaining straps on these immense craft were, much to Hoverlloyd's consternation, regularly breaking in service. Engineers in Bradford schemed a resilient coupling in rubber and nylon tyre cord and this 'Bonio' device proved far more durable in use than its high-tensile steel predecessor. In another fine example of Bradford's quality of manufacture, in 1969 a routine inspection of an electric multiple unit by British Rail turned up a Spencer Moulton rubber brake hose that was overdue for replacement. There was nothing unusual in this, of course, as such hoses were fitted to almost all trains and were replaced every five years – what was surprising was that this particular hose had been in service for over thirty years and, whilst showing no sign of failure, was replaced on a precautionary basis.

Despite the march of progress, the Kingston Mills site itself remained broadly similar, continuing its expansion onto every spare patch of ground as production requirements and staff numbers rose inexorably – as had been the pattern since the 1850s. A hotch-potch of buildings contained all manner of machinery from the latest injection moulding machines to museum pieces such as the 'Iron Duke' calender. In the early 1970s there were 1,200 people (with around 900 on the factory floor) working at 'The Avon' – as it had come to be known – and the site, bounded by the River Avon to the south, the town centre to the west and The Hall to the north, was quite full. With yet more manufacturing capacity required and the option of new buildings now off the menu, productivity and efficiency improvements were needed; as well as new facilities, new machinery and new working practices. Bradford's future lay in modernisation; and its landscape was to change forever.

The main entrance to the works – Kingston House on right, the Gatehouse in the centre and the Kingston Mill dominating the skyline in the late 1960s. (WSA)

10   Moulding and Modernization

"Many jobs could be threatened if expansion plans are thwarted."
Tony Mitchard, GM at Bradford

It is perhaps misleading to state that new buildings were no longer possible in Bradford. It was perfectly feasible to erect new buildings provided that space was made available for them by removing existing ones. And so it came to pass that in 1972 an application was made to Wiltshire County Council to demolish the Kingston Mill itself and replace it with a modern 66,000 square foot, two storey workshop – 'Shop 30'. The Mill, built by Thomas Divett in 1805, was originally used for the weaving of cloth. As with many of Bradford's mills, it was abandoned in 1842. It will be remembered that Stephen Moulton established his rubber factory in the Kingston Mill in 1849 and it had been in continuous use since that date. However, as working methods and machinery changed over time, the Mill had become less useful.

In the early days, the water wheels were used to provide power to the factory, but these had long been usurped by steam and later augmented by electricity. As recently as 1970 the six huge Lancashire coal-fired boilers had been replaced with two oil-fired burners capable of producing 20,000 lbs of steam per hour. The working area was spread over five storeys, limiting what work could be undertaken on the upper floors. Despite strong opposition, permission was granted for demolition of the mill which had been described as 'one of the most beautiful cloth mills in England'. Those who acquiesced in this were accused of civic irresponsibility, but were in fact taking the pragmatic view that the factory needed to modernize and if denied the facility to do so would be forced to move elsewhere, taking Bradford's livelihood with it. General Manager at Bradford, Tony Mitchard, cautioned that the Kingston Mill "did not offer practical production capability" and that "many jobs could be threatened if expansion plans are thwarted." The two bells that Thomas Divett had installed in the Belfry on the roof of the Mill to call the workers in to the factory were gifted to Christ Church in the town, and the Belfry itself was given to Alex Moulton. The water turbine from the mill was preserved and put on display at the Bowerhill Industrial Estate in Melksham. Stephen Moulton's 1849 calendar machine, the 'Iron Duke' was dismantled and placed in the care of the Bristol Industrial Museum.

Another casualty of the age was the name of George Spencer, Moulton & Co. With the production at Bradford increasingly blurred between Avon group companies, the axe finally fell on Spencer Moulton in Avon's 1974 reorganisation. The Bradford site

One of the finest cloth mills in England - Kingston Mill in the 1960s. (WSA)

The Kingston Mills site, viewed from the top of Abbey Mill in the early 1970s. Modern buildings are beginning to appear on this site, as Avon Rubber chased increased productivity and output. The Kingston Mill can be seen in the centre, with its distinctive belfry cupola. (WSA)

Kate Blackmore and Natalie Galley sorting and trimming aerosol cup seal blanks. Avon in Bradford provided up to 95% of the worldwide requirements for aerosol gaskets and cup seals. (WSA)

Derek Middleton moulding hose parts for Hoover, using a modern 'Desma' moulding machine. (WSA)

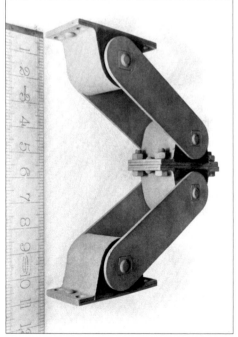

Spencer Moulton Flexitor gangway (carriage interconnection) spring. Still to be found on most trains passing through Bradford on Avon (and elsewhere, of course). (AMCT)

became the principal factory for Avon Industrial Polymers Ltd. – 'Rubber' presumably sounded too prosaic for the bright new world of the 1970s. Far from relegating Bradford to a small division of a large conglomerate, this move reinforced its prime position in the group as developer and manufacturer of advanced rubber products. Avon Industrial Polymers (AIP) employed a total of 2,335 staff of which 1,300 were based in Bradford – the remainder were based at Melksham (650), Birmingham (340) and at Tessenderlo in Belgium (46).

AIP had six major product groups at the time, these being Automotive, Technical Products, Domestic Appliances, Railways, Dairy & Agriculture, and Sports Equipment. The latter two categories were located away from Bradford, although there was always the possibility of moving manufacture to utilise capacity where it was available. In addition, most 'Avon Group' functions were carried out in the Abbey Mill, including central services, personnel, printing, engineering and patents. Although most of the Abbey Mill had been given over to offices by this time, aerosol gaskets were still manufactured on the ground floor in enormous quantities. In 1973 no fewer than twenty million seals were produced each week, with 95% of aerosols worldwide using AIP (Bradford) gaskets.

The Automotive products were primarily moulded hoses for cooling and heating systems. These were manufactured in Bradford and AIP supplied approximately 60% of the original equipment hoses used in British cars and a successful export drive led to Volkswagen ordering no fewer than three million hoses in 1977. Home-grown offerings such as the Ford Cortina initially adopted AIP dust-covers for ball-joints, a small part but significant business as Dagenham manufactured 1,400 Cortinas every day. British Leyland called off over seven million AIP parts every year for the Morris Marina alone. Also of note was the persistent demand for Flexitor suspension units for use on Bramber trailers.

Flexitor units also still featured in the railway business, principally as gangway springs between carriages – indeed, even today, the modern trains that pass through Bradford sport these between each carriage. The recently introduced British Rail 'Mk2' coach featured no less than 100 Avon parts that were manufactured in Bradford. Of course the railway business spread further than rolling stock; and despite the three-day week working in 1974 (Bradford worked Monday to Wednesday) Avon supplied 2.25 million rail pads to Taiwan. The Domestic Appliance operation was largely Bradford based, and was chiefly concerned with supplying hoses – an offshoot from the automotive hose business.

The energy crisis brought new challenges – as well as power, the price of rubber rose to levels not seen since the late 1940s, and rayon and nylon used in tyres were rationed. Coupled with rampant inflation, this hit Avon hard. Chairman Hugh Rogers outlined the plan for the future: "husband our cash resources, eliminate uneconomic operations and back our proven successes." Despite an increase in turnover to 72 million, in 1975 Avon posted a loss of 0.76 million. John Swanborough resigned, as did Finance Director Bryan Horton. Avon Industrial Products felt the full force of Rogers' words – the factories in Tessenderlo and Birmingham were closed and profitable contracts were moved to Bradford.

Recovery, however, was swift. Avon made a profit of 2.5 million in 1976 and 5 million the year after. In 1977, with capacity at Bradford unable to cope with burgeoning demand running at 300,000 hoses a week, a new 105,000 square foot hose factory was built in Trowbridge to replace the existing 43,000 square foot Bradford facility. This was soon followed by a new factory in Chippenham to handle 'flexible fabrications'. Such was the pressure on production that the Chippenham factory was doubled in size almost immediately and the original Melksham site was significantly expanded. Even with this expansion in production facilities, AIP still reported capacity issues at Bradford, despite the startling statistic that the Abbey Mills were manufacturing no less than fifty million aerosol gaskets every week.

Pressure was relieved by a new factory for Avon Industrial Polymers in Chippenham in 1979. This gave a little breathing space and saw the establishment of a Custom Moulding Department in Bradford under Ian Hastie. Working on the 'sprat to catch a mackerel' principle, this new facility looked to solve unusual problems in a cost-effective manner and draw new business into volume manufacturing. Early successes included the use of scrap rubber to weigh down traffic cones, mudflaps for the Ford Motor Co., rubber skittle balls and moulded shafts for fireman's axes. The axes were particularly practical as they protected against electric shock; they were to remain in production at Bradford until the bitter end.

The spectre of recession loomed again in the early 1980s and the slender profit in 1980 tipped into a more significant loss in 1981. In a programme of general rationalisation and retrenchment, the Bradford site of Avon Industrial Products lost its autonomy as it merged with the Melksham site. Despite this change, when the redundancy axe inevitably fell in 1982, it was the Melksham site that suffered with 250 workers laid off. Chairman Peter Fisher re-affirmed that Avon Industrial Polymers at Bradford, Chippenham, Melksham and Trowbridge

Testing the viscosity of rubber. Close control over raw material and compounding was essential to ensure the quality of the finished product.

had not escaped the effects of recession but were trading profitably. The tyre business, however, was clearly losing money with the result being the loss of 600 jobs and the closure of the Bridgend plant.

The Kingston Mills site continued to deal with the more unusual and esoteric requests from customers both old and new. The expansion of the North Sea Oil business brought orders for five-and-a-half tonne bumper pads for oil rigs and a £300,000 contract for 'Avonclad' anti-fouling polymer coatings – 3,500 square metres of these cupro-nickel impregnated neoprene sheets were used on each oil rig. The automotive industry requirements continued to increase, with Volkswagen, Volvo and Rover being the prime purchasers – the Rover 200 alone used 2.5 million Avon rubber parts each year. Demand increased further with the opening of the Nissan factory in Sunderland. Avon invested £1.5 million in a major efficiency programme at Bradford which saw the introduction of Computer Aided Design and fully-automated, computer controlled moulding machines. Stock control and work flow was computerised with subsequent increase in on-time deliveries and drastic reduction of stock levels. Notwithstanding this, traditional manufacture still remained, particularly in the case of railway buffer springs.

Modern automated 'Desma' moulding machines (WSA)

Right up until the end, Avon Rubber were vulcanizing buffer springs in the traditional manner in Bradford on Avon. (WSA)

It is worth pausing a moment here to take stock of the position of the Kingston Mills rubber factory in Bradford. Since taking over in 1956, Avon Rubber had re-organised and re-equipped the site to face the challenges of the modern business world. Simple compression moulds had given way to automated injection- and transfer-moulding. Capacity had been greatly increased and buildings expanded and modernized. The majority of management functions had been transferred from Melksham to Bradford and the technical reputation of Bradford staff was second to none. The centre of power remained in Manvers House on Kingston Road; the top brass occupied the lower floors whilst the Product Designers worked on the top floors. Tooling design was based in the new offices adjacent to Manvers House, whilst the Centenary Building opposite continued to fulfil its original intended function as a laboratory. Quality and Personnel departments were based in Kingston House. Beyond this lay the five-acre factory site, nearly 300,000 square feet of production space enclosed within a jumble of buildings and rooflines old and new packed into the compact site bounded by the Kingston Road, the River Avon and the railway. The Abbey Mills, downstream of the Town Bridge, continued to manufacture die-cut aerosol gaskets, now at a rate of 80 million per week – 132 every second and some 45% of the world's requirements. Most of the town's workers earned their living at 'The Avon', as had been the case for 140 years. In the latter part of the 20th century, they numbered over 1,000.

The New Mills (right) and the Vaults (ahead) – the Lamb Yard and the 'Black Hole'.
Contrast this with how the Lamb Yard looks today! (WSA)

Avon's annual turnover punched through the £200 million barrier in the late 1980s. Profit levels, however, remained weak and as the UK dipped into recession in 1991 Avon announced another round of redundancies – 300 jobs were to go (out of a total workforce of 3,200) in Wiltshire. Concerns grew as to where these jobs would be lost; these were later to be confirmed as 90 in Bradford, 40 in Chippenham, 30 at Melksham and 20 in Trowbridge – and another 120 in the tyre division in Melksham. Avon made a further commitment to the Kingston Mills site with the announcement of a very substantial investment into a modern rubber mixing plant at Bradford, to be installed by the end of 1992. More ominous was the comment from Chief Executive Tony Mitchard that "the Bradford on Avon factory has been in a serious loss-making situation for some time now." With rumours growing as to the future of the Kingston Mills site, Avon were moved to officially deny any planned closure in mid 1991.

'The Avon' at full capacity in the 1990s – from the Town Bridge in the foreground to the Mould Store (Lower Greenland Mill) in the far top right, factory buildings cover the whole site. Note The Hall at centre top. The Kingston Mill has gone, but the red-brick rubber factory buildings from the 1890s remain. To the right, the swimming pool has been demolished but the library has not yet been built. (MBC)

By October the devastating truth was out – Avon were to close the Kingston Mills in its entirety by the following June, and to cease production at the Abbey Mills site by the end of 1992. This was presented as a 'fait accompli' to the workers, although the blow was softened somewhat by generous voluntary redundancy terms and no enforced redundancy. However, no amount of 'softening' and platitudes from management could take away the feeling of great loss that seemed, to some, as if the heart had been ripped out of Bradford on Avon. Many people were angry - employees, union officials, locals - but, in reality, there was very little that they, or anyone else, could do. The site where, in 1848 Stephen Moulton had founded his rubber factory, manufactured the capes for the Crimean War and equipped the railways of the British Empire, was to fall silent. The Mills where generations of families had utilised their skills and earned their livelihoods were to close, to be demolished, and to be re-developed. The clutch of local businesses that relied on 'The Avon' for the bulk of their work would be badly – often fatally – affected; the mould makers, ironmongers and general engineers. The everyday traders would see the loss of over 1,000 people past their doors every day – the grocers, butchers and, of course, the public houses. Several were to shut up shop forever as business became unviable.

As the remaining staff contemplated the logistics of commuting to Chippenham every day – somewhat of a change for most, who had for years walked to the Kingston Mills – the closure of the site played out in slow-motion. Windscreen wiper blade manufacture moved out to Trowbridge, and the remainder of the automotive business moved to Chippenham. Railway mechanicals re-located to Spencer Moulton France, another Avon subsidiary. Several products were sold off, others were dropped entirely. In some cases, where the production machinery was unviable to move, enormous quantities were manufactured before the equipment was abandoned. Gradually, the lights went out across the site as each department closed. Compounding and mixing were amongst the last to go. Given that the bulk of Bradford manufacture was relocated into a modern 80,000 square foot factory – less than one third of the Kingston Mills site – it was, and indeed still is, hard to argue with the promised efficiency gains and the overall business strategy.

A 'wake' was held in Chippenham for the Kingston Mills site on June 1st 1992. For some, it was the first day of their new jobs away from Bradford; for others, their last day before retirement or a change of direction. With the move complete, site clearance began. Led by Avon Services Project Manager Eddie Hook, this team searched the basements and attic storage areas across the site and unearthed bundles of hand-drawn illustrations and engineering drawings, some of which dated from the 1850s. One building alone contained well over 100 plan chest drawers. Other items included wooden casting patterns, watchmen's clocks, antique workshop tools and an unopened tin containing three unused Spencer Moulton tennis balls. Amongst all of this was the formula for Stephen Moulton's patented 'Hypo Mixture':

Take 2 1/2 lb of caustic soda and put it in a vessel with 5 gallons of clear water. Let it be together alone about 10 to 12 hours (occasionally stirring). Then add 25 lb of sulphur and 85 lb of litharge, stir it morning and night for the first week, after that period stir once a day. In about one month it will be ready to wash. Take it and wash three times in clean water, afterwards dry it and it is fit for use. For stronger Hypo add more sulphur. June 16th, 1854.

Despite Avon's stated intent, the Abbey Mills remained open beyond 1992, punching out millions of aerosol gaskets every day. As it turned out, its reprieve was merely a stay of execution and Avon's manufacturing operations in Bradford closed their doors for the last time in 1995.

The last shipment from the Kingston Mills – one of the mixing machines, craned onto a trailer, is moved out along Kingston Road in 1993.
The building in the background is Kingston House. (AS)

Spencer Moulton
SPORTS BALLS

*T*he Spencer Moulton is the ideal INDIA rubber ball for Physical Training, Gymnastic and Rhythmic exercises, Sports, Games and Pastimes. Made in 4 sizes, 2½, 3, 4 and 5 inches diameter and a range of delightful colours.

Epilogue

Following the cessation of rubber manufacture at the Kingston Mills, attention soon turned to how the site would be developed – indeed the Mayor of the town, Peter Taylor (an Avon employee himself) addressed this problem as soon as closure was announced: "It would be a disaster if it stayed for years and deteriorated, but hopefully that won't happen. Bradford is a nice place to come to." It is worth noting that, unusually, the factory was located right in the centre of the town, with all the benefits and downsides that such a position brings. Amongst the former is the attractiveness of the town itself; for the latter we have the problems of traffic, access and river flooding. Avon had hoped to sell the site quickly, and the town was keen to see a new use for the site, but both were to be disappointed. Whilst the Abbey Mill was re-developed (into retirement flats) soon after closure in 1995, the Kingston Mills site was to descend into a state of dereliction over a period of nearly twenty years.

A surprising result of this delay was that almost as soon as Avon left the town, they came back. With the Kingston Mills empty and unsold, Avon made Manvers House their Head Office and it held that position from 1995 until the sale of the site in 2005 and indeed, although now let to others, Avon still retain ownership. By that time many of the other buildings, including the Lamb Factory and Kingston House, were in a lamentable condition. In 2010, most of the factory buildings were demolished to clear the way for housing development. It is gratifying to see that the significant buildings that remain on the site have been restored to their former glory. A few years ago, however, aside from the rubber tree carved in stone on the front of the Centenary Building, it was hard to find any trace of Bradford's once world-renowned rubber industry.

Thanks to the efforts of the Bradford on Avon Museum and Bradford on Avon Preservation Trust, the 'Iron Duke' calender machine has been restored, re-assembled and returned to the town. On permanent display outside Kingston House, the 'Iron Duke' is a fitting memorial to rubber's contribution to Bradford's industrial history, and to the generations of men and women who worked at the Kingston Mills.

One could be forgiven for thinking that, following Avon's departure, rubber vulcanization no longer takes place in Bradford. You would be mistaken, for in a corner of the stable block in the grounds of The Hall lies a small rubber moulding press. The stable block is part of the Moulton bicycle factory, and this humble machine produces an array of rubber springs and bump-stops for Alex Moulton's iconic small-wheeled bicycle which is still manufactured here in Bradford on Avon. The rubber is mixed elsewhere, but the vulcanization process itself takes place right here in the town. You will not find automation or computer controls on this press, rather simple transfer and compression moulding in metal moulds that Stephen Moulton himself would recognize – although he would be astonished at the rubber engineering made possible by rubber to metal bonding and the design of torsional shear springs.

Another unusual remnant of Bradford's rubber industry may be found in central France. One may recall that Spencer Moulton established several overseas offices and factories in the early part of the twentieth century. Most of these were acquired by Avon Rubber in 1956 and many were re-named, re-purposed, sold or simply closed down. The French facility, however, retained its title and was kept under the Avon umbrella until as late as 2003 – long after the demise of the Spencer Moulton name in England. Now in management ownership, Spencer Moulton S.A. still design and manufacture railway mechanicals – notably buffer springs – in a similar fashion to George Spencer's and Stephen Moulton's businesses in the nineteenth century.

Moulton Bicycle Co.'s Rubber Moulding Press (DF)

Epitaph

Verse engraved on Spencer Moulton buffer,
author unknown (photo: DF)

My task should be, but for the idiosyncrasies of the B.T.C.,
With wagons hauling coal and ore,
Absorbing loads of eighty tons or more.
Now I'm banished to Alex Moulton's study floor,
A jilted lover – ousted by a damned hydraulic buffer.
Why can't he present me to Mayfair's Bagatelle,
To be a bar stool for some Bond Street Jezebel?

Rubbery Rot.

In quiet little Bradford,
On Avon, not in Yorks,
In bye-streets, and at bridge end,
Stand Spencer-Moulton's Works.

These Works are very famous.
They make a lot of things,
Doorsteps, tyres, and crutch-pads,
And Spencer's Buffer Springs.

Their staff is very learned,
They work from morn till night.
Mr. Q is ever busy
Seeing things are right.

Their Chemical Department
Is doubtless of the best,
And Doctor P. and Mr. T.
Are up to any test.

No Works can well exist without
Some Drawings and a plan.
The D.O. is the place for this,
And Mr. M. their man.

Requests for them to function
Are to Mr. B. addressed,
Who enters them up in a book.
The Works should do the rest !

The cost of making many things,
Some curious persons ask.
The producing of the answers
Is Mr. D's hard task.

The superhuman energies
Of members of the firm
Are noted down by Mr. B.,
Who tots up what they earn.

The inspection of the products
Is carried out with care,
In fact, the careful Mr. L.
Makes some producers swear.

" Correspondence *must* be governed " !
And this is done *so* well.
The hiding place of letters
Is the secret of Miss L.

Their output and their energy
Are entered up each week.
Miss M. you go and interview,
If such details you'd seek.

The controlling of this mighty staff,
A hard task you may guess,
Is done, extraordinarily well,
By their able Mr. S.

Rubber manufacture is
A long and tedious task,
For instruction in the mixing,
Their Mr. G. you ask.

Mr. H. he sees it weighed
And washed for Messrs. SS.,
Who, by mixing it and rolling it,
Produce some sticky messes.

For " Pleasant Railway Travelling "
Their springs are quite the rage.
Made by another Mr. S.,
They last for quite an age.

For railway work or gardens,
They make a perfect pipe.
Mr. W. makes it for them,
When he isn't catching pike !

Is this then not your size of tyre ?
It's twenty-eight by three.
They most strongly recommend it.
It was made by Mr. V.

The despatching of the articles
They entrust to Abbey Mills,
Where Mr. F. and Mr. T.
Are busy with their bills.

This little song it now must end.
Admittedly it's rot !
But when other folk are working,—
The writers ? They are not.

BARDS OF BRADFORD-ON-AVON.

'Rubbery Rot', first published in the 'Winged Arrow', Vol. 1, No. 3, 1920

The Rubber Industry's Buildings

KINGSTON HOUSE

The mansion that is now called Kingston House stands on the southern side of Kingston Road (formerly - before 1890 - Mill Street). It has had various other names- Mill Street House, Rivers, Avon House and the prosaic 'Shop 55' of the Kingston Mills Rubber Works. Confusingly, it is not the only building in the immediate area to have used the title 'Kingston House' – the Jacobean mansion known as 'The Hall' was known by this name until the mid-nineteenth century.

It seems to have begun as a small Palladian house in the early 18th century, with industrial premises alongside.

A house here was rated in 1727 and by the 1740s was occupied by Robert Knight, the first of a series of clothiers who included, or may have included, the Methuens, Rogers, Hilliers and Cams. In the 1790s it, and Manvers House across the road, belonged to Thomas Bush and it was bought in 1841 by wool-dyer Charles Spackman. His children sold it in 1869 to the clothiers James Harper and Thomas Taylor, along with neighbouring New Mills.

After a very brief spell as a girls' school the George Spencer, Moulton Company bought it in 1899 and used it as offices until the succeeding Avon Company abandoned it in 1992. During World War 2 it served as the headquarters of Bradford's Local Defence Volunteers (the Home Guard). The plaque affixed to the front of the building commemorates those from the rubber factory who gave their lives in the First, and Second, World War.

Along with most of the Kingston Mills site, Kingston House fell into a state of dereliction after being vacated by Avon Rubber in 1992. As part of the re-development of the site the building has been returned to its former glory, but at the time of writing is not occupied. The small gatehouse to the west of Kingston House has also been restored. (Photo MUS)

GREENLAND MILLS

- Greenland Upper Mill c.1804 (John Hinton, later Thomas Tugwell)
- Greenland Middle Mill c.1807 (Stoddart, Gale, Howell & Co)
- Greenland Lower Mill c.1808 (William and Philip Shrapnell)

Greenland's development came in the 18th century with the industrialisation of woollen cloth manufacture and a great factory building (Greenland Upper Mill) was in existence by 1804. The Middle and Lower Mills were built by 1808. The Greenland Mills were operated by Thomas Divett who, with partners, had purchased much of the old Hall estate. The greatest development came after 1851 at the hands of J.W. Applegate & Co, who extended the mill buildings with acres of mechanised weaving factories that surrounded both Greenland Upper and Greenland Middle Mills.

Woollen cloth ceased to be made in 1905 and the buildings were put to other uses- initially they were occupied by the Sirdar Rubber Company. Sirdar went bankrupt and their facilities were taken over by the Avon India Rubber Co. in 1915. When Avon left the principal establishment was Dotesio's printing works. Later incumbents included the Rex Rubber Company, MY Sports, Marcos cars and Weir Electrical Instruments.

The Lower Greenland Mill, lying at the north-western end of the weir, was truncated and cut-off from the river by the new railway line which crosses the Avon and the millstream at this point. Around 1855 it was purchased by Stephen Moulton and it became a workshop for the carpenters on The Hall Estate. After the Second World War it was sold (along with the rest of the Kingston Mills and Paddock site) to Spencer, Moulton & Co. and was used as a storeroom for rubber moulding tools. This building still stands (empty), in a tight corner between railway and millstream right at the far (eastern) end of the new Kingston Mills development.

During the 1980s the companies working in the Upper and Middle Mills had to leave in advance of development of the whole site for housing, which did not happen until the increasingly derelict and listed Upper Mill was burnt down by arsonists. Since then, a replica of the old factory has been built as a block of flats and surrounded by houses. (Photo DF)

KINGSTON MILL

Kingston Mill was built for Thomas Divett, a London Factor, around 1805. Divett also owned The Hall, the grandest house in Bradford, and in true West-country style the Mill was positioned so that it was overlooked by The Hall. Kingston Mill survived an early arson attempt by weavers protesting against the mechanisation of their work. By 1816 Divett had let it to Hopkins and Howard, 'Manufacturers of Luxury and Superfine Cloths' who occupied it until 1826. In 1836 it was let to Samuel Pitman. His business failed (as did many others in the town) by 1842 and the Mill remained empty until bought, along with The Hall, by Stephen Moulton in 1848.

Standing a full five storeys (and roof space) high and described as 'one of the most beautiful cloth mills in England', the

Kingston Mill was the centre-piece of the Moulton Rubber factory. Modern production methods eventually made the building obsolete and it was demolished in 1972. No trace remains on the site, but the water turbine from the Mill can be found at the south end of the Bowerhill Industrial Estate in Melksham. (Photo AMCT/DF)

MANVERS HOUSE

Manvers House is named after the Earls Manvers. The first Earl Manvers, Charles Meadows, was awarded the title in 1806. Charles was a nephew of the second Duke of Kingston who succeeded to the Hall estate after 1788. Charles died in 1816. The second Earl Manvers died in 1860, the third in 1900 and the fourth in 1926.

Clothier Matthew Smith built Manvers House on the north side of the road, perhaps in the 1730s. Samuel Cam, a clothier, living at the Chantry House in the town, bought Manvers House from the estate of Matthew Smith. Cam's purchase was to set up his daughter and his son-in-law Isaac Hillier, also a clothier.

Thomas Bush, another clothier, bought Isaac Hillier's property in Mill Street, including Manvers House and the associated workshops. He died in 1809 aged 68 and the Hillier property on the north side of Mill Street passed to his son John, an attorney. John lived at Woolley Hill House in 1841, held many public roles in the town, including that of a Town Commissioner, and had an office in Church Street. At that date the plots 402 workshops, 403 garden, 404 house and yard, 405 gardens and 416 house, all in Mill Street (which included Manvers House), were let by him to Benjamin Matthews. He was one of the most prosperous traders at Bradford, a saddler and harness maker and also an innkeeper.

Manvers House is probably the house in Mill Street which was occupied by N. Jarvis Highmore, M.D. in 1865. In 1911 Charles Edward Stewart Flemming, physician and surgeon, lived at Manvers House. He was one of two medical officers and public vaccinator for nos. 1 and 2a districts of Bradford Union. In 1939 the house was occupied by Doctors Charles Flemming and A.A.G. Flemming (the latter presumably being the formers son).

In 1940 Manvers House was bought by George Spencer, Moulton & Co. For the duration of the War, their Head Office was re-located here and the Chairman, Jack Spencer, and his wife lived in an apartment upstairs. Later the building was extended and used solely for commercial purposes. Being separated from the Kingston Mill site by Kingston Road, Manvers House was not vacated by Avon Rubber when the site closed, and indeed Avon's Head Office was situated in Manvers House from 1995 to 2005. Whilst leased to other companies (currently Hitachi), Avon still own the building.

The house was listed grade II* in 1952. (Photo DF)

NEW MILLS and THE VAULTS: 'The Black Hole'

The mid-19th century building on the north-western edge of the site was originally used for weaving, and was probably associated with the New Mills woollen mill directly to the south-east. The New Mills were built at a similar time, initially as a rectangular four-storey building with single storey extension of similar size. Around 1869 the height of the extension was raised to the same height as the original building and a three storey building was attached at the southern end. George Spencer, Moulton & Co. bought the New Mills from Harper and Taylor clothiers in or around 1898, and the single storey weaving shed shortly afterwards. An additional storey was added to the shed, and the ground floor was converted, by the provision of stone-arched vaults, to a fire-proof store for raw rubber stocks. The adjacent New Mills building was utilised for rubber mixing and compounding. This messy, dirty job led to the whole complex being known locally as the 'black hole'.

The insurance plan from 1908 shows the group of buildings comprising the Vaults (the former weaving shed), the New Mills and other sundry buildings at the west end of Spencer-Moulton's property being labelled as the 'Lamb Factory' as distinct from the 'Kingston Mills' factory to the east, centred around the Kingston Mill itself. Later plans include all the western buildings under the heading of 'Kingston Mills'. Both the Vaults and the New Mills were in use right up until the end in 1992 - compounding and mixing were the last processes to leave.

Despite old photographs showing the apparent disrepair of these buildings, much of the external and internal structure survived in good order. These buildings have been restored as part of the redevelopment of the site, and are now a mix of residential, retail and hospitality units. To provide a much-needed additional entrance to the site, a large hole has been punched through the southern end of the New Mills building to allow a road through - this practical expedient does not detract from the setting. (Photo DF)

LAMB BUILDING (BUILDING 70), KINGSTON MILLS, SILVER STREET

The Lamb building is located at the north east side of the town bridge. Its name is inherited from the Bradford Brewery Company's public house, 'The Lamb Inn' that was on this site until purchased and demolished by Spencer Moulton in 1916. Wishing to utilise this new space available to them as much as possible, the Lamb building was originally schemed to be four storeys high – perhaps to mimic the Abbey Mill downstream on the other side of the bridge.

Built around 1916 to a revised design of only two storeys, the Lamb building is unusual in that it was one of the first in Britain to be constructed with a reinforced concrete frame. The architect was E.J.C. Manico. The concrete frame was manufactured in London by the Trussed Concrete Steel Construction Company and erected on site under their direction. As well as being quick and cheap to build, reinforced concrete buildings are durable and resistant to vibration and fire – all of these being desirable qualities for a factory building. Despite this, concrete construction was slow to become accepted in Britain and the Lamb building is a rare early example.

Principally used for the manufacture of hoses and rubber flooring, the Lamb building now houses a restaurant and a small supermarket. Renovation works uncovered substantial problems with the original foundations and extensive restoration and rebuilding was required to preserve this substantial and unusual building on the townscape. (Photo DF)

CENTENARY BUILDING / RESEARCH LABORATORY

Erected in 1948 to mark the Centenary of the establishment of Stephen Moulton's rubber company, this was, in effect, built to house Spencer Moulton's new Research Department. The timing was coincident with the closure of the Technical Office in London.

Alex Moulton persuaded the board that, as well as being functional, this should be a 'significant' building and he was allowed to choose (and work with) the architect. His choice was GN Gordon Hake, principal of the West of England School of Architecture. Situated immediately facing Manvers House, the resulting stone-faced three storey building managed to respect the proportions and features of its earlier counterpart whilst being modernist in design and execution.

Dr Pickles' laboratory and office was located on the ground floor. Alex Moulton's office was strategically positioned on the top floor, overlooking the factory and with a view eastward to The Hall. A stylised version of the company's logo - depicting a man tapping the latex from the rubber tree - is featured in stone on the front of the building.

Perhaps surprisingly, this building is not listed but has been restored as part of the redevelopment of the Kingston Mills site and is now in residential use. (Photo DF)

ABBEY MILL

The present Abbey Mill, built in 1875, replaced an earlier building that dated from 1807 or earlier. It was the last purpose-built mill to be built in the town. Square-set and five storeys high, it is built right on the north bank of the river Avon and dominates the view westwards from the town bridge. Designed by Richard Gane of Trowbridge, it shares some features with many of the Trowbridge Mills that were built by the Gane family.

Originally used by Harper and Taylor, then Harper, Taylor and Ward, followed by Ward, Taylor and Willis, Abbey Mill was offered for sale in 1898. The sale particulars detail some 14 mules (each with between 300 and 400 spindles) and 16 scribbling engines in the mill. For some years after this, the Abbey Mill was used for rug manufacture.

From 1915 the mill was being used by Spencer Moulton for the manufacture of rubber goods. Latterly, under the ownership of Avon Rubber, the upper floors of Abbey Mill were converted into open-plan offices and a staff restaurant. The ground floor was used for manufacture of gaskets and cup seals for aerosols – at one time some 80% of the world market for these was supplied from the Abbey Mill.

Following closure in 1995, the Abbey Mill was converted into retirement flats.
(Photo DF)

GRIST MILL

The Grist Mill that stands in the middle of the Kingston Mills site dates from the late 18th century, although it sits on the site of a medieval mill which may have existed since the eleventh century. Much of what can be seen today is more recent, as most of the 18th century structure was badly damaged by fire in 1901.

The Grist Mill was bought by Thomas Divett in 1805 and was substantially rebuilt at the same time that the neighbouring Kingston Mill was constructed. After being let to Hopkins and Howard (along with the Kingston Mill) in 1820 it was sold by Divett to Stephen Moulton in 1848. After this date it was used by Moulton for the manufacture of rubber goods.

Following the fire in 1901 the Grist Mill was rebuilt on the same footprint but to a greater height. Re-development of the site following cessation of rubber manufacture saw the Grist Mill being retained for architectural and historical interest. Modern re-working for residential use included extending the whole building to three storeys and the addition of a pitched roof behind a parapet. (Photo DF)

THE HALL

This magnificent house, the largest in Bradford on Avon, is set on the eastern side of the town close to the site of the medieval mill and upstream of the 'broad ford' on the Avon after which the town is named. It was originally known as The Hall, then called the Great House, then either Kingston House or The Duke's House, then became again The Hall. It is listed Grade 1 and acknowledged to be of national importance architecturally.

The estate has been owned by four families. For centuries it was the seat of the Hall family, who held it as a sub-manor of Shaftesbury Abbey's large manor of Bradford. In the 18th century it was owned by the Duke of Kingston (Earl Manvers) family, followed some forty years later with the Divett family in industrial use. From 1848 onwards it has belonged to the Moulton family.

The house has been considered a classic example of early Jacobean architecture. But there was an earlier house on the site and recent research has shown that in about 1610-20, instead of a completely new build, the existing manor house was extensively remodelled. Fashionable new work was cleverly grafted onto the pre-existing structure and the whole raised in height resulting in an ornamental rural 'lodge' of the kind in vogue at the time, able to sustain entertainment on a lavish scale for short periods.

Though at its height the estate was quite sizeable, amassed through a series of good marriages, it was never huge. Only the immediate surroundings were left with the house in 1805 but after 1848 the Moulton family gradually acquired more property in Bradford, most of which has since been resold. Following the passing of Dr. Alex Moulton in 2012, The Hall and its remaining pocket of grounds has been cared for by the Alex Moulton Charitable Trust. (Photo DF)

Appendix: Charles Goodyear

The first chapters of this book gave some detail on how Charles Goodyear discovered the rubber vulcanisation process and his attempts to sell the rights to his invention in England. After Hancock had proved his patent in the courts at the expense of Moulton and Goodyear, the latter gentleman falls outside of the scope of this book. However, Goodyear's story does not end there. As it is of more than passing interest it is well worthy of recording here.

Following on from Goodyear's defeat in the English courts, he returned to America. Rubber was in the ascendancy again, and Goodyear's knowledge was in demand. He also held US Patent 3633 – 'Improvement in India-Rubber Fabrics' – the key vulcanisation patent. So keen was he to make some money – largely to pay off debts – Goodyear made a string of ill-advised licence agreements that paid him a little or nothing. Furthermore, to protect his invention he had to defend it at the US Supreme Court no less than 32 times. Nevertheless, he did make some money. Goodyear, who had once been jailed for non-payment of a $5 hotel bill, was able to pay Secretary of State Daniel Webster $15,000 to represent him in another infringement case – the largest fee ever paid to an American lawyer at the time. The court found in his favour, but it didn't stop people using his process without his consent.

Goodyear saw rubber as the material of the future. He wanted to make everything out of it: banknotes, musical instruments, flags, jewellery, sails, even the ships themselves. He had his portrait painted on rubber, his business cards engraved on it, his autobiography printed on and bound in it. He wore rubber hats, vests, and ties. He ate off rubber plates and drank from rubber beakers. As he found new uses, he sold his ideas for a handful of dollars and went back to his workshop. He flirted persistently and dangerously with bankruptcy and was jailed for non-payment of debts on several occasions – he referred to the debtor's jail as his familiar hotel.

Goodyear fathered twelve children, tragically six died in infancy. For years his family relied on the benevolence of others to eke out an existence whilst he pursued his experiments. Whilst some of his contemporaries made fortunes from his discoveries, Goodyear himself saw little return on his lifetime's work. One could forgive him for harbouring regrets, but he was more than magnanimous in his reflection on his efforts. "Life," he wrote, "should not be estimated exclusively by the standard of dollars and cents. I am not disposed to complain that I have planted and others have gathered the fruits. A man has cause for regret only when he sows and no one reaps."

He died in New York in 1860, at the age of 59. He left debts totalling over $200,000.

Bibliography

Woodruff, William, *The Rise of the British Rubber Industry*, Liverpool UP 1958
Payne, Peter, *Rubber and Railways in the nineteenth century*, Liverpool UP 1961
Dobson, Margaret, *Bradford Voices*, Ex Libris Press 2011
Loadman, John, *Tears of the Tree*, Oxford UP 2014
Slack, Charles, *Noble Ambition*, Hyperion, 2003
George Spencer, Moulton & Co., *One Hundred Years of Rubber Manufacture*, Spencer Moulton 1948
George Spencer, Moulton & Co., *Bradford Rubber Products go all over the World*, Wiltshire News reprint 1952
George Spencer, Moulton & Co., *Six Vital Years*, Spencer Moulton 1946
Avon Rubber plc, *Avon 1885-1985,* Avon Rubber 1985
Moulton, Alexander, *Bristol-to-Bradford on Avon*, Rolls-Royce Heritage Trust 2009
Bardsley, Gillian, *Issigonis: The Official Biography*, Icon 2005
Slocombe, Pamela, *The Hall, Bradford on Avon*, Ex Libris Press 2012
Ponting, K. G. *The Hall*, Not published
Gazard, David, *Abbey Mill*, Bradford Museum
Wood, Jonathan, *Alec Issigonis: The Man who Made the Mini*, Breedon 2005
Demidowicz, George & Toni, *Kingston Mills*, Bradford-on-Avon 1999
Nash Partnership, Kingston Mills Listed Building Statement/Historic Building Record 2006

On-line resources:

The Victoria History of the County of Wiltshire
Grace's Guide
Bradford-on-Avon Museum
http://iisrp.com   Brief History & Introduction of Rubber
www.freshford.com
britishhistory.ac.uk: India Rubber

Archives and Records:

George Spencer, Moulton & Co. by courtesy of The Wiltshire and Swindon History Centre, Chippenham
The Avon Rubber Company by courtesy of The Wiltshire and Swindon History Centre, Chippenham
Archives of the Late Dr Alex Moulton by courtesy of The Alex Moulton Charitable Trust

Map

Wiltshire XXXII (includes: Atworth; Bradford-on-Avon; Holt; Monkton Farleigh; South Wraxall.)
Surveyed 1886; Published 1889

About the author

Dan Farrell has been immersed in the world of Moulton bicycles for over twenty years. His involvement covers design, engineering, manufacture, and riding – notably completing the 1200km Paris-Brest-Paris randonneé in 1995 and 2003. Educated at Brunel University – a first degree in Industrial Design and a second in Design Management – Dan is both a Chartered Engineer and a Chartered Technological Product Designer. He is currently Technical Lead at the Moulton Bicycle Company in Bradford on Avon and also works at Pashley Cycles and as a professional consulting engineer. Dan is a councillor and trustee of the Institution of Engineering Designers. He has lived in Bradford on Avon since 2014.

Acknowledgements

Thanks to: Kate Berry, Richard Cook, Steve Missen, Mervyn Harris, Ivor Slocombe, Guy Vincent, Pamela Slocombe, Gillian Marston, Andrew Shipley, Margaret Shipley, Michael Griggs, Roger Clark, Roger Jones, Julian Orbach, Tony Hadland.

Photo caption key

AMCT	Alex Moulton Charitable Trust
GSM	George Spencer Moulton
MUS	Bradford on Avon Museum
WBR	Wiltshire Buildings Record
DTR	DTR UK, Trowbridge
WSA	Wiltshire and Swindon Archives
DL	Dave Lee
DF	Dan Farrell

Publication of this book has been made possible with the benefit of grants from the Heritage Lottery Fund, the Alex Moulton Charitable Trust and Moulton Bicycle Company